Understanding Remote Sensing

Understanding Remote Sensing

Ryan Fox

Published by States Academic Press,
109 South 5th Street,
Brooklyn, NY 11249, USA
www.statesacademicpress.com

Understanding Remote Sensing
Ryan Fox

© 2022 States Academic Press

International Standard Book Number: 978-1-63989-545-8 (Hardback)

This book contains information obtained from authentic and highly regarded sources. All chapters are published with permission under the Creative Commons Attribution Share Alike License or equivalent. A wide variety of references are listed. Permissions and sources are indicated; for detailed attributions, please refer to the permissions page. Reasonable efforts have been made to publish reliable data and information, but the authors, editors and publisher cannot assume any responsibility for the validity of all materials or the consequences of their use.

Trademark Notice: Registered trademark of products or corporate names are used only for explanation and identification without intent to infringe.

Cataloging-in-Publication Data

Understanding remote sensing / Ryan Fox.
   p. cm.
Includes bibliographical references and index.
ISBN 978-1-63989-545-8
1. Remote sensing. 2. Aerial photogrammetry. 3. Aerospace telemetry.
4. Space optics. I. Fox, Ryan.

G70.4 .U53 2022
621.367 8--dc23

TABLE OF CONTENTS

Preface VII

Chapter 1 **Remote Sensing and its Types**..1
- Microwave Remote Sensing..11
- Thermal Remote Sensing...19
- Hyperspectral Remote Sensing...28

Chapter 2 **Remote Sensing and Image Sensing**..38
- Aerial Photography and Photogrammetry......................................38
- Photogeology..54
- Platforms and Sensors..61
- Image Rectifications..74
- Image Enhancement..79
- Image Transformation..95

Chapter 3 **RADAR and LiDAR**..102
- RADAR...102
- Types of RADAR..111
- LiDAR...128
- Types of LiDAR..133

Chapter 4 **GIS and GPS**...139
- Geographic Information System...139
- Global Positioning System...157
- Google Earth...190

Chapter 5 **Uses of Remote Sensing**..207
- Watershed..207
- Runoff Model..215
- Irrigation Management...222
- Flood Mapping...224
- Environmental Monitoring..227

Permissions

Index

PREFACE

The science of gathering information about a distant object without making any physical contact with it is called remote sensing. It makes use of different kinds of sensors to observe the Earth and other planetary bodies. Remote sensing measures the radiation emitted and reflected by an area to analyze its physical characteristics. The sensors can be classified into active and passive sensors. Active sensors respond to internal stimuli while the passive sensors use external stimuli. They record the natural energy which is reflected from the Earth's surface. The important characteristics of data which is collected are spatial resolution, spectral resolution, radiometric resolution and temporal resolution. The field of remote sensing finds extensive application in the fields of geography, land surveying, intelligence, economics and commercial planning. The topics included in this book on remote sensing are of utmost significance and bound to provide incredible insights to readers. Different approaches, evaluations, and methodologies of remote sensing have been included herein. This book is a complete source of knowledge on the present status of this important field.

To facilitate a deeper understanding of the contents of this book a short introduction of every chapter is written below:

Chapter 1- The process of scanning and recording the physical properties of an object or phenomena without making physical contact with the object is known as remote sensing. It is done by measuring the object's reflected and emitted radiations. There are various types of remote sensing such as microwave remote sensing, thermal remote sensing and hyperspectral remote sensing. This is an introductory chapter which will introduce briefly all the significant aspects of remote sensing.

Chapter 2- Photogrammetry involves the techniques used to obtain information of real world objects through recording, measuring and interpreting photographic images. Taking photographs from an aircraft or other flying object is known as aerial photography. This chapter discusses in detail all aspects related to remote sensing and image processing like photogeology, image rectifications, image enhancement etc.

Chapter 3- RADAR is Radio Detection And Ranging while LiDAR is Light Detection And Ranging. Radar is an electromagnetic sensor which uses radiowaves to determine the angle, range and velocity of objects. Lidar is a remote sensing method used to measure distances by using light in the form of pulsed laser. The aim of this chapter is to explore the various types of RADAR and LiDAR.

Chapter 4- A computer system used for recording, storing, examining and displaying data related to positions on Earth's surface is known as Geographic Information System. Global Positioning System is a navigation system that provides geolocation and time information by using satellites, a receiver and algorithms. All the diverse principles of Geographic Information System have been carefully analyzed in this chapter.

Chapter 5- Remote sensing has emerged as an effective tool for analysis and better management of natural resources. It has applications in various fields such as irrigation management, flood mapping, environmental monitoring, runoff model etc. The diverse applications of remote sensing in the current scenario have been thoroughly discussed in this chapter.

I owe the completion of this book to the never-ending support of my family, who supported me throughout the project.

Ryan Fox

Chapter 1
Remote Sensing and its Types

The process of scanning and recording the physical properties of an object or phenomena without making physical contact with the object is known as remote sensing. It is done by measuring the object's reflected and emitted radiations. There are various types of remote sensing such as microwave remote sensing, thermal remote sensing and hyperspectral remote sensing. This is an introductory chapter which will introduce briefly all the significant aspects of remote sensing.

Remote sensing is a method of collecting information about any ground object under investigation from a distance without being in contact. There are many definitions found, however, the most accepted definition was given by the American Society for Photogrammetry and Remote Sensing in 1988. Accordingly, Remote Sensing can be defined as "The art, science and technology of obtaining reliable information about physical objects and the environment, through the process of recording, measuring and interpreting imagery and digital representations of energy patterns derived from non-contact sensor systems."

Advantages and Limitations of Remote Sensing

Remote Sensing techniques have several advantages over the conventional field based investigations. These are:

- Synoptic Overview: The remote sensing images provide a synoptic overview or bird's eye view of a larger area, enabling us to study the relationship among different ground objects and delineation of regional features/trends.

- Feasibility Aspects: Due to inaccessibility to ground survey in many parts of the terrain, remote sensing is the only scientific method for data collection.

- Time Saving: Remote sensing saves time and manpower as larger area can be covered by this technique.

- Unobtrusiveness: If the remote sensors collect the information passively by recording the electromagnetic energy reflected or emitted by the ground object, the area of interest is not disturbed. It also ensures collection of information in its natural state.

- Systematic Data Collection: Remote Sensing devices collect the information of the ground surface in a systematic manner with a specific time interval, removing the sampling bias introduced in some in situ investigations.

- Derivation of Biophysical Data: Under controlled conditions, remote sensing can provide fundamental biophysical information, e.g., location, elevation, temperature, moisture content, etc.

- Multi-disciplinary Applications: The same remote sensing data may be used by researchers or workers from different disciplines, e.g., geology, forestry, agriculture, hydrology, planning, defense, etc. and therefore, increase the overall benefit-to-cost ratio.

Although, there are many advantages making the technique an enormously popular tool, it has some limitations too:

- Understanding limit of application: The greatest limitation of this technique is that its utility is often oversold. Remote Sensing alone cannot provide all the information needed for any scientific study. The applicability of these tools and techniques are limited to selection of appropriate sensors, its resolutions, time of data collection and appropriate post-processing operations.

- Expensive technique: The collection and interpretation of remote sensing data is expensive, as it requires specific instrumentation and skills. However, the enormous advantages of this technique overrule this limitation.

Basic Principles of Remote Sensing

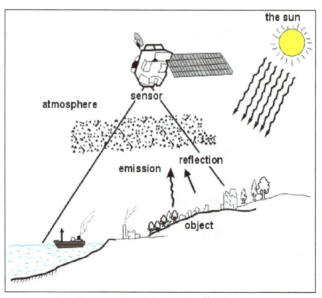

Remote Sensing Data Collection.

All objects of the earth's surface (at 300 Kelvin) like, soil, rock, vegetation, etc. above absolute zero (-273 °Centigrade or 0 Kelvin) emit electromagnetic energy. And so does the Sun (at 6000 Kelvin). Sun is the major source of energy required for remote sensing purpose (except radar and sonar). The energy is transferred by electromagnetic radiation through the vacuum between the Sun and the Earth at the speed of light. It interacts with the atmosphere before coming into contact with the earth's surface. While returning, it interacts with the atmosphere once again and finally reaches the remote sensor. The detectors or photographic film system on board records this reflected or emitted energy in analogue or digital form.

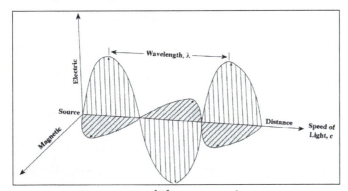

Components of Electromagnetic wave.

The electromagnetic radiation principle conceptualized by James Clerk Maxwell in 1960, refers to all energy that moves with the velocity of light (300,000 km per second) in a harmonic wave pattern. The electromagnetic wave consists of two fluctuation fields – one electrical and the other magnetic at the right angle to one another. Both are also perpendicular to the direction of travel. The electromagnetic waves are characterized by its wavelength (i.e. the distance from any point on one cycle or wave to the same position on the next cycle or wave measured in micrometer, µm) and frequency (number of cycles or waves pass through a point per second). The frequency is inversely proportional to the wavelength.

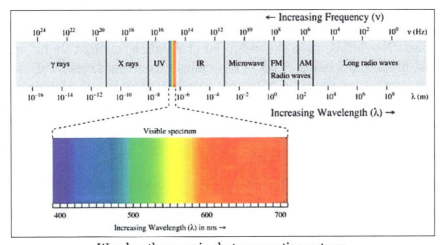

Wavelength ranges in electromagnetic spectrum.

The basic principle involved in remote sensing methods is that in different wavelength ranges of the electromagnetic spectrum, each type of object reflects or emits a certain intensity of light, which is dependent upon the physical or compositional attributes of the object. Hence, the spectral behavior (i.e. the intensity of light emitted or reflected by the objects) of same ground object in different wavelength ranges may be studied through Spectral Signature Curves. Such curves may help to differentiate different types of objects, e.g., soil, vegetation, water body, settlements, etc. and map their distribution on the ground. The remote sensing missions are, thus, a process of collection of spectral information of the ground objects, enhancement and interpretation for different applications.

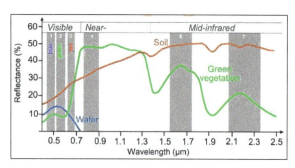 

Spectral Signature Curves of common ground objects. Spectral bands of Landsat-7 are shown in the background (1 µm = 1000 nanometers = 10-6 meters).

Four bands of IRS LISS-3 multi-spectral image showing different spectral characteristics of various ground objects.

A Typical Remote Sensing Process

Statement of Problem

Remote sensing can provide information on various biophysical (viz. location, elevation, chlorophyll concentration, biomass density, surface temperature, soil moisture, evapotranspiration, snow/ice cover, etc.) and hybrid (viz. land use, vegetation stress, etc.) variables. Depending upon the nature of problem, a scientist should identify the potential use of remote sensing. Scale of mapping and accuracy specification should be given the priority during selection of appropriate sensor and methodology.

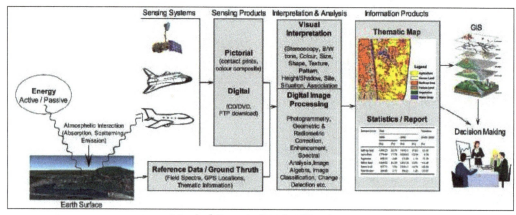

A typical Remote Sensing Program.

Data Acquisition

Remote sensing data may be collected using either passive or active remote sensing systems. Passive sensors record naturally occurring electromagnetic radiation that is reflected or emitted from the terrain. Remote sensing in the day light under the influence of solar energy falls under this category. When man-made electromagnetic energy is used to illuminate the ground and backscatters are recorded by the sensor (e.g. in microwave radar), such sensors are called passive sensors.

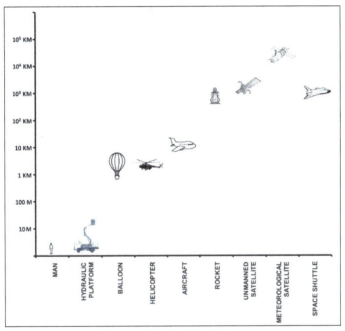

Remote Sensing Platforms.

Remote sensing images are collected from suitable platforms located at various altitudes e.g. aerial (balloons, helicopters and aircraft) and space-borne (rockets, manned and unmanned satellites). Hydraulic platforms and handheld spectroradiometers are used to generate ground truth data. Each remote sensing system is characterized by four types of resolutions e.g. spectral, spatial, temporal and radiometric. Thorough understanding of these resolutions is needed to extract meaningful information from remote sensing data.

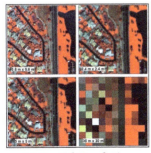
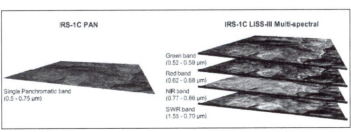

(a) Spatial Resolution.   (b) Spectral Resolution.

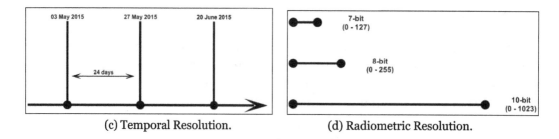

(c) Temporal Resolution.	(d) Radiometric Resolution.

Data Processing, Analysis and Interpretation

Remote Sensing data acquired through various sensors and platforms are processed to enhance the quality for better analysis and interpretation. Aerial photographs mostly available in analogue form are meant for visual interpretation (through tone, texture, pattern, association, etc.) and photogrammetric measurements. However, the satellite remote sensing data collected in digital form can be used for quality enhancement, statistical and syntactical pattern recognition, photogrammetric processing, expert system and neural network image analysis. The image processing tasks are carried out in computer workstations with the help of specific image processing software (e.g., Erdas Imagine, ENVI, eCognition, Idrisi/TerrSet). Through these processes, various thematic information may be extracted e.g. land cover, lithology, structure, soil, vegetation cover, etc.

Field/Ground Validation

Field data or ground truth plays very important roles in remote sensing. These are:

- Calibration of remote sensor.
- Data correction, analysis and interpretation.
- Accuracy assessment of thematic maps generated from remote sensing data.

Timing of field data collection should be decided based on the nature of application. The time-stable parameters, such as, spectral emissivity, rock type, structure, etc. may be collected any time, however, time variant parameters (viz. temperature, rain, condition of crop, phonological cycle) must be measured during the remote sensing overpass. In case of archival data analysis, fieldwork is usually carried out in the same season for identical weather or phonological condition. In most of the cases, purposive sampling strategy is adopted utilizing skills and local knowledge of the field worker. In the field, thematic information e.g., lithology, structures, landform, plant species, soil types, water bodies, etc. are collected that help to verify the remote sensing image interpretations. Location information for observation points are collected through GPS instruments. In specific cases, field spectroradiometers are also used to generate spectral signature curves of ground objects that may be used to identify similar objects in the remote sensing data. Recently, very high spatial resolution Google Earth images have also been used as reference data for interpreting relatively low resolution remote sensing data for regional scale applications (e.g., Land use/land cover mapping, etc.)

Presentation for Decision Making

The outputs are presented as analogue or digital thematic maps, spatial database file, statistic or graphs. The final outputs often become an important data source for GIS-based decision support system. Most recently, integrated approaches of GIS and remote sensing have become more effective tools than using remote sensing interpretation alone.

Interpretation is the processes of detection, identification, description and assessment of significant of an object and pattern imaged. The method of interpretation may be either visual or digital or combination of both. Both the interpretation techniques have merits and demerits and even after the digital analysis the output are also visually analysed.

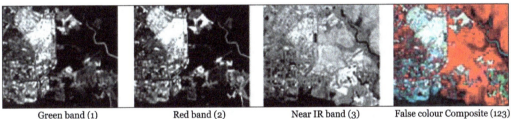

Green band (1)   Red band (2)   Near IR band (3)   False colour Composite (123)

Combination of 3 bands generates colour composite images.

The ability of human to identify an object through the data content in an image/photo by combining several elements of interpretation. There are two types of extraction of information from the images/photographs namely:

- Interpretation of data by visual analysis.

- Semi-automatic processing by computer followed by visual analysis likes generation of vector layer from raster image through onscreen digitization and DTM/DEM generation. Similarly interpretation of aerial photographs through 3D generation through visual studies. In general analog format in remote sensing data is being used in visual interpretation. This involves the systematic examination of data, studying existing maps, collection of field information and works at various levels of complexity. The analysis depends upon the individual perception, and experience of the interpreter, nature of the object, quality of the data, scale, combination of special bands etc.

The entire process of visual interpretation can be divided into following few steps namely detection of an object, interpretation, recognition and identification, analysis, classification, deduction and idealisation and based on this identifying an object conclusion. Hence interpretation is the combined result of identification of feature through photo recognition elements, field verification and preparation of final thematic maps. It also requires the process of observation coupled with imagination and great deal of patience.

Basic Elements of Interpretation

The interpretation of satellite imagery and aerial photographs involves the study of various basic characters of an object with reference to spectral bands which is useful in visual analysis. The basic elements are shape, size, pattern, tone, texture, shadows, location, association and resolution.

- Shape: The external form, outline or configuration of the object. This includes natural features Man Made feature.

- Size: This property depends on the scale and resolution of the image/photo. Smaller feature will be easily indented in large scale image/photo.

- Pattern: Spatial arrangement of an object into distinctive recurring forms: This can be easily explained through the pattern of a road and railway line. Even though both look linear, major roads associated with steep curves and many intersections with minor road.

- Shadow: Indicates the outline of an object and its length which is useful is measuring the height of an object. The shadow effect in Radar images is due to look angle and slope of the terrain. Taller features cast larger shadows than shorter features.

- Tone: Refers to the colour or relative brightness of an object. The tonal variation is due to the reflection, emittance, transmission or absorption character of objects. This may vary from one object to another and also changes with reference to different bands. In General smooth surface tends to have high reflectance, rougher surface less reflectance. This phenomenon can be easily explained through Infrared and Radar imagery.

- Infrared imagery: Healthy vegetation reflects Infrared radiation much stronger than green energy and appears very bright in the image. A simple example is the appearance of light tone by vegetation species and dark tone by water. Particularly in thermal infrared images the brightness tone represents warmest temperature and darkness represent coolest temperature. The image illustrates daytime and night time thermal data. The changes in kinetic water temperature cause for the tonal changes. Hence time is also to be taken consideration before interpretation.

- Radar Imagery: Smooth surfaces reflect highly and area blocked from radar signal and appears dark. Bridges and cities show very bright tone, on the contrary calm water, pavement and dry lake beds appear very dark tone.

- Texture: The frequency of tonal change. It creak a visual impression of surface roughness or smoothness of objects. This property depends upon the size, shape, pattern and shadow.

- Location Site: The relationship of feature to the surrounding features provides clues to words its identity. Example: certain tree species words associated with high altitude areas.

- Resolution: It depends upon the photographic/imaging device namely cameras or sensors. This includes of spectral and spatial resolutions. The spectral resolution helps in identifying the feature in specific spectral bands. The high spatial resolutions imagery/photographs are useful in identifying small objects.

- Association: Occurrence of features in relation to others.

Issues in Interpretation

- Unfamiliar scale and resolutions.
- Lack of understanding of physics of remote sensing.
- Understanding proper spectral character of each object.
- Visually interpret 3 layers of information at a time.

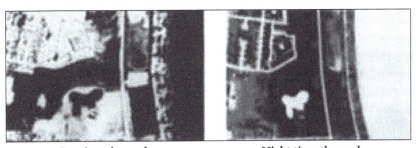

Daytime thermal.   Night time thermal.

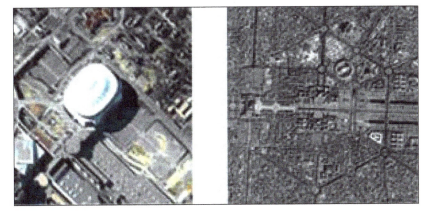

Success of Interpretation

- Training and Experience of the interpreter.
- Quality of photo/Images.
- Local knowledge of the study area.

Advantages in Visual Interpretation

- Simple method.
- Inexpensive equipment.
- Uses brightness and spatial content of the image.
- Subjective and Qualitative.
- Concrete.

Advantages of Digital Image Processing

- Cost-effective for large geographic areas.
- Cost-effective for repetitive interpretations.
- Cost-effective for standard image formats.
- Consistent results.
- Simultaneous interpretations of several channels.
- Complex interpretation algorithms possible.
- Speed may be an advantage.
- Explore alternatives.
- Compatible with other digital data.

Disadvantages in Digital Processing

- Expensive for small areas.
- Expensive for one-time interpretations.
- Start-up costs may be high.
- Requires elaborate, single-purpose equipment.
- Accuracy may be difficult to evaluate.
- Requires standard image formats.

- Data may be expensive, or not available.
- Preprocessing may be required.
- May require large support staff.

High Resolution Images

The spatial resolution of an image /photograph is one of the main elements to decide the scale of map and classification level of information. The advances in sensor development led the acquiring of high resolution imagery which helps planners, and professionals for making large scale maps, better planning and monitoring. The panchromatic data of IKONOS with the resolution of 1mt/4mt is being used widely for large scale mapping. These high resolution imagery can help in fast visual interpretation by human eyes, the best image processor of all and even small objects be easily detected. Hence the human brain can identify an object by understanding, analyzing the context. The chances of misinterpretation of object are very less in visual analysis because of the expertise, experience and local knowledge of the field.

Microwave Remote Sensing

The microwave region of the electromagnetic spectrum extends from wavelengths of about 1 mm to about 1 m and is divided in bands based on the wavelength as shown in Table. This region is, far removed from those in and near the visible spectrum, where our direct sensory experience can assist in interpretation of images and data.

Table: Bands in Microwave region of the EM spectrum.

Designation	Frequency range	Wavelength range
L band	1 to 2 GHz	15 cm to 30 cm
S band	2 to 4 GHz	7.5 cm to 15 cm
C band	4 to 8 GHz	3.75 cm to 7.5 cm
X band	8 to 12 GHz	25 mm to 37.5 mm
Ku band	12 to 18 GHz	16.7 mm to 25 mm
K band	18 to 26.5 GHz	11.3 mm to 16.7 mm
Ka band	26.5 to 40 GHz	5.0 mm to 11.3 mm

The microwave remote sensing has many advantages compared to optical remote sensing such as:

- Microwaves penetrate the atmosphere through clouds and rain as the longer wavelengths are not susceptible to atmospheric scattering.

- Detection of microwave energy is possible under almost all weather and environmental conditions.

- Day and night operation as independent of the sun as source of illumination.

- Penetration depth into vegetation and soil and subsurface penetration.

- Allows accurate measurements of distance.

Active and Passive Microwave Remote Sensors

Active Microwave Sensors

Active microwave sensors provide their own source of microwave radiation to illuminate the target. It consists fundamentally of a transmitter, a receiver, an antenna, and an electronics system to process and record the data. Thus, an active microwave sensor broadcasts a directed pattern of energy to illuminate a portion of the Earth's surface, then receives the portion scattered back to the instrument. This energy forms the basis for the imagery we interpret. Active sensors generate their own energy, so their use is subject to fewer constraints, and they can be used under a wider range of operational conditions. Further, because active sensors use energy generated by the sensor itself, its properties are known in detail. Therefore, it is possible to compare transmitted energy with received energy to judge with more precision than is possible with passive sensors the characteristics of the surfaces that have scattered the energy. Active microwave sensors are generally divided into two distinct categories: imaging and non-imaging.

Imaging Radar

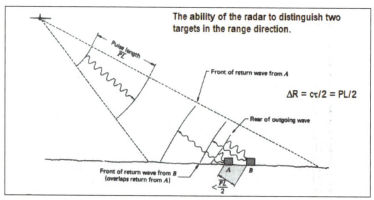

Range resolution is dependent on pulse length.

Imaging radar is similar to a photograph taken by a camera, but the image is of radar waves, not visible light. The most common form of imaging active microwave sensors is RADAR. RADAR is an acronym for Radio Detection and Ranging. The sensor transmits a microwave (radio) signal towards the target and detects the backscattered portion of the signal. The strength of the backscattered signal is measured to discriminate between different targets and the time delay between the transmitted and reflected signals determines the distance (or range) to the target.

Transmitter generates successive short bursts (or pulses of microwave (A) at regular intervals which are focused by the antenna into a beam (B). The radar beam illuminates the surface obliquely at a right angle to the motion of the platform. The antenna receives a portion of the transmitted energy reflected (or backscattered) from various objects within the illuminated beam (C). By measuring the time delay between the transmission of a pulse and the reception of the backscattered "echo" from different targets, their distance from the radar and thus their location can be determined. As the sensor platform moves forward, recording and processing of the backscattered signals builds up a two-dimensional image of the surface. Since, the energy in air propagates at approximately the velocity of light, the slant range, SR is equal to ct/2, where c is speed of light, t is the time between pulse transmission and echo response. For the radar to distinguish between two objects, the spatial resolution should be equal or greater the half the pulse length as shown in figure above. Even though the SR does not change with the distance of the transmitter, the ground-range (GR) resolution varies. As shown in figure, GR becomes smaller with the increase in SR range. Thus, $GR = \dfrac{c\tau}{2\cos\theta_d}$, where τ is pulse duration, θ_d is the depression angle.

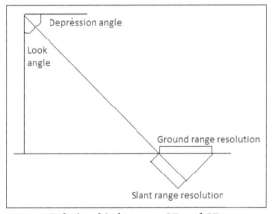

Relationship between GR and SR.

The imaging radar finds its application in distinguishing the sea ice as it typically reflects more of the radar energy emitted by the sensor than the surrounding ocean, which makes it easy to distinguish between the two. However, the amount and character of reflected energy are functions of the physical properties of the sea ice and thus, can be difficult to interpret radar images of sea ice. In general, though, thicker multiyear ice is readily distinguishable from younger, thinner ice because radar energy bounces back to the sensor from the bubbles in the ice left when brine drains. This feature makes synthetic aperture radar (SAR), shown in Figure, an especially useful tool for measuring the extent of thick vs. thin sea ice. E.g. The RADARSAT mission, managed by the Canadian Space Agency provides with detailed images of sea ice. SAR employs a short physical antenna and through utilization of techniques, it synthesizes the effect of a very long antenna. So here a single physical antenna is linked into array linked antennas. Thus, SAR map variations in microwave backscatter at fine spatial scales (10 to 50 m), and is used to create an image, which measures the variations in surface roughness and surface moisture.

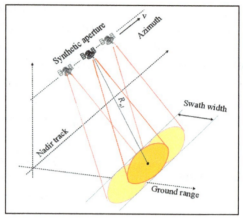

Illustration of SAR.

Non-Imaging Radar

Non-imaging radar is profiling devices which take measurements in one linear dimension, as opposed to the two-dimensional representation of imaging sensors. The example instruments for non-imaging radar are Scatterometer and altimeters. Scatterometer makes precise quantitative measurements of the amount of energy backscattered from targets. The amount of energy backscattered is dependent on the surface properties (roughness) and the angle at which the microwave energy strikes the target. Ground-based scatterometers are used extensively to accurately measure the backscatter from various targets in order to characterize different materials and surface types. The Sea-Winds sensor aboard NASA's Quick Scatterometer (QuikSCAT) satellite provides daily, global views of ocean winds and sea ice. Altimetry sensor sends a pulse of radar energy toward the earth and measures the time it takes to return to the sensor. The pulse's round-trip time determines how far the satellite is from the reflecting surface. Altimeters look straight down at nadir below the platform and thus measure height or elevation. With a known reference, this information is used to measure the altitude of various features at the earth's surface. With enough precision, a radar altimeter can determine the height of the sea ice surface above sea level, which is utilized to calculate the total thickness of the sea ice. Thus, non-imaging radar finds application in ocean topography, glacial ice topography, sea ice characteristics and classification of ice edge.

Passive Microwave Sensors

A passive microwave sensor detects the naturally emitted microwave energy within its field of view. Passive microwave sensors don't rely on reflected sunlight as with the passive optical sensors. Thus, they don't need to be placed in sun-synchronous orbits so they can be placed on almost any platform. Passive microwave sensors receive the emitted energy in microwave region of EM spectrum from the objects on the ground. The Objects at the earth's surface emit not only infrared radiation; they also emit microwaves at relatively low energy levels. The microwave energy recorded by a passive sensor can be emitted by the atmosphere (1), reflected from the surface (2), emitted from the surface (3), or transmitted from the subsurface (4). Example of Passive microwave sensors - Radiometers. Since, the wavelengths are so long, the energy available is quite small compared to optical wavelengths. Thus, the fields of view must be large to detect enough energy to record a signal.

Most passive microwave sensors are therefore characterized by low spatial resolution. The process of the passive type is explained using the theory of radiative transfer based on the law of Rayleigh Jeans, which in the microwave region gives, $T_b = \varepsilon_\lambda T$, where Tb is the brightness temperature and ε, is the emissivity of the object. Emissivity is the emitting ability of a real material compared to that of a black body that varies with composition of material and geometric configuration of the surface. It is a ratio and varies between 0 and 1 and for most natural materials it ranges between 0.7 and 0.95. In both active and passive types, the sensor may be designed considering the optimum frequency needed for the objects to be observed. In passive microwave remote sensing, the characteristics of an object can be detected from the relationship between the received power and the physical characteristics of the object such as attenuation and/or radiation characteristics.

Applications of Microwave Remote Sensing

Microwave remote sensing finds its application in various fields a given below:

- Meteorological applications: Allows measuring the atmospheric profiles and determine water and ozone content in the atmosphere.

- Hydrological applications: Measure soil moisture since microwave emission is influenced by moisture content. So microwave remote sensing can be utilized for oceanographic applications like mapping sea ice, currents, and surface winds detection of pollutants, such as oil slicks.

Various satellites are available, which utilize the microwave remote sensing:

- AQUARIUS: It is a combination of radiometer and radar. Radar measures winds for correcting for the effect of surface roughness. It was the first instrument to measure global ocean salinity. Passive and active microwave instrument operate at L-band, with a Resolution - Baseline 100km, Minimum 200km. The global coverage is in 8 days and the accuracy is 0.2 psu.

- TOPEX/Poseidon and Jason-1: It is a joint NASA-CNES Program, where TOPEX/Poseidon launched on August 10, 1992 and Jason-1 was launched on

December 7, 2001. It consists of Ku-band and C-band dual frequency altimeter. It also has microwave radiometer to measure water vapor, GPS, DORIS, and laser reflector for precise orbit determination. The sea-level measurement accuracy is 4.2 cm.

- SRTM: It has C-band single pass interferometric SAR for topographic measurements using a 60m mast. It has been useful for providing DEM of 80% of the Earth's surface in a single 11 day shuttle flight. It covered 60 degrees north and 56 degrees south latitude at 57 degrees inclination. The current best estimate of the SRTM accuracy is 10 m horizontal and 8 m vertical.

- ESA's Soil Moisture Ocean Salinity (SMOS): Earth Explorer mission is a radio telescope in orbit, which has Microwave Imaging Radiometer using Aperture Synthesis (MIRAS) radiometer. It picks up faint microwave emissions from Earth's surface to map levels of land soil moisture and ocean salinity. The satellite captures images of 'brightness temperature', which correspond to radiation emitted from Earth's surface.

- NASA's Soil Moisture Active Passive satellite can provide accurate coarse resolution soil moisture information at 40 km. Measurements from the Copernicus Sentinel-1 satellite can then be applied to improve the resolution to 'field scale'. By combining measurements from different sensors the spatial resolution is increased from 40 km to 100 m.

- CloudSAT: CloudSAT measures the vertical structure of clouds and quantify their ice and water content. It helps in improving weather prediction and clarifies climatic processes. The mission has been able to investigate the way aerosols affect clouds and precipitation. It also allows the utility of 94 GHz radar to observe and quantify precipitation in the context of cloud properties.

Microwave Scattering

Microwave surface scattering is the process by which microwave radiation incident upon a solid or liquid surface is wholly or partially redirected away from that surface. Microwave emission is the process by which microwave radiation originates from a solid or liquid surface due to the rotational or vibrational motions of molecules near the surface.

Scattering Parameters

For a two-port network, as shown in the following figure, if the power is applied at one port, most of the power escapes from the other port, while some of it reflects back to the same port. In the following figure, if V_1 or V_2 is applied, then I_1 or I_2 current flows respectively.

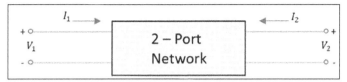

Structure of a two port network.

If the source is applied to the opposite port, another two combinations are to be considered. So, for a two-port network, 2 × 2 = 4 combinations are likely to occur. The travelling waves with associated powers when scatter out through the ports, the Microwave junction can be defined by S-Parameters or Scattering Parameters, which are represented in a matrix form, called as "Scattering Matrix".

Scattering Matrix

It is a square matrix which gives all the combinations of power relationships between the various input and output ports of a Microwave junction. The elements of this matrix are called "Scattering Coefficients" or "Scattering S Parameters". Consider the following figure:

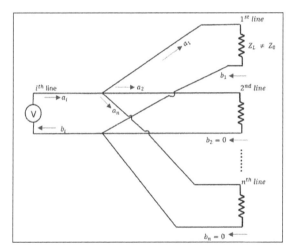

Here, the source is connected through i^{th} line while a_1 is the incident wave and b_1 is the reflected wave. If a relation is given between b_1 and a_1^1,

$$b_1 = (reflection\ coefficient)a_1 = S_{1i}a_1$$

Where;

- S_{1i} = Reflection coefficient of 1st line where i is the input port and 1 is the output port.
- 1 = Reflection from 1st line.
- i = Source connected at i^{th} line.

If the impedance matches, then the power gets transferred to the load. Unlikely, if the load impedance doesn't match with the characteristic impedance. Then, the reflection occurs. That means, reflection occurs if,

$$Z_l \neq Z_o$$

However, if this mismatch is there for more than one port, example 'n' ports, then $i = 1$ to n since i can be any line from 1 to n. Therefore, we have,

$$b_1 = S_{11}a_1 + S_{12}a_2 + S_{13}a_3 + \ldots\ldots\ldots + S_{1n}a_n$$
$$b_2 = S_{21}a_1 + S_{22}a_2 + S_{23}a_3 + \ldots\ldots\ldots + S_{2n}a_n$$
$$\vdots$$
$$b_n = S_{n1}a_1 + S_{n2}a_2 + S_{n3}a_3 + \ldots\ldots\ldots + S_{nn}a_n$$

When this whole thing is kept in a matrix form,

$$\begin{bmatrix} b_1 \\ b_2 \\ b_3 \\ \cdot \\ \cdot \\ \cdot \\ b_n \end{bmatrix} = \begin{bmatrix} S_{11} & S_{12} & S_{13} & \cdots & S_{1n} \\ S_{21} & S_{22} & S_{23} & \cdots & S_{2n} \\ \cdot & \cdot & \cdot & \cdots & \cdot \\ \cdot & \cdot & \cdot & \cdots & \cdot \\ \cdot & \cdot & \cdot & \cdots & \cdot \\ S_{n1} & S_{n2} & S_{n3} & \cdots & S_{nn} \end{bmatrix} \times \begin{bmatrix} a_1 \\ a_2 \\ a_3 \\ \cdot \\ \cdot \\ \cdot \\ a_n \end{bmatrix}$$

Column matrix $[b]$ Scattering matrix $[S]$ Matrix $[a]$

The column matrix $[b]$ corresponds to the reflected waves or the output, while the matrix $[a]$ corresponds to the incident waves or the input. The scattering column matrix $[s]$ which is of the order of n × n contains the reflection coefficients and transmission coefficients. Therefore,

$$[b] = [S][a]$$

Properties of [S] Matrix

The scattering matrix is indicated as $[S]$ matrix. There are few standard properties for $[S]$ matrix. They are:

- $[S]$ is always a square matrix of order $n \times n$ $[S]_{n \times n}$.

- $[S]$ is a symmetric matrix i.e $S_{ij} = S_{ji}$.

- $[S]$ is a unitary matrix i.e., $[S][S]^* = I$.

- The sum of the products of each term of any row or column multiplied by the complex conjugate of the corresponding terms of any other row or column is zero. i.e.:

$$\sum_{i=j}^{n} S_{ik} S_{ik}^* = 0 \text{ for } k \neq j$$

$(k = 1, 2, 3, ... n)$ and $(j = 1, 2, 3, ... n)$

- If the electrical distance between some k^{th} port and the junction is $\beta_k I_k$, then the coefficients of S_{ij} involving k, will be multiplied by the factor $e^{-j\beta k l k}$.

Thermal Remote Sensing

The word 'thermal' is pertaining to heat or temperature and in thermal remote sensing, heat radiated by the imaged surface is the main source of data. Thus thermal remote sensing is defined as the branch of remote sensing that deals with the acquisition, processing and interpretation of data acquired primarily in the thermal infrared (TIR) region of the electromagnetic (EM) spectrum. Most commonly used spectrum is the intervals from 3 to 5 µm and 8 to 14 µm, in which the atmosphere is fairly transparent and the signal is only lightly affected by atmospheric absorption. In Thermal remote sensing, we are interested that how well energy is emitted from the surface at different wavelengths. Thermal remote sensing does not depend on reflected sunlight, so that it can also perform during the night.

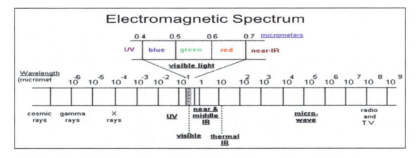

It is a well-known fact that all natural targets (features) reflect as well as emit radiations. In the TIR region of the EM spectrum, the radiations emitted by the earth due to its thermal state are far more intense than the solar reflected radiations and therefore, sensors operating in this wavelength region primarily detect thermal radiative properties of the ground material. As thermal remote sensing deals with the measurement of emitted radiations, for high temperature phenomenon, the realm of thermal remote sensing broadens to encompass not only the TIR but also the short wave infrared (SWIR), near infrared (NIR) and in extreme cases even the visible region of the EM spectrum.

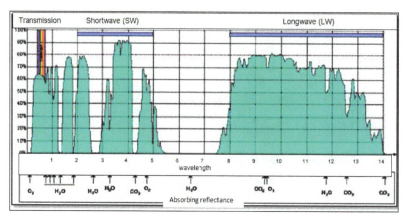

Basic Thermal Radiation Principles

Spectral Emissivity and Kinetic Temperature

In thermal remote sensing, radiations emitted by ground objects are measured for temperature estimation. These measurements give the radiant temperature of a body which depends on two factors; kinetic temperature and emissivity. Thermal remote sensing exploits the fact that everything above absolute zero (0 K or -273.15 °C or −459 °F) emits radiation in the infrared range of the electromagnetic spectrum. How much energy is radiated, and at which wavelengths, depends on the emissivity of the surface and on its kinetic temperature.

Emissivity is the emitting ability of a real material compared to that of a black body and is a spectral property that varies with composition of material and geometric configuration of the surface. Emissivity denoted by epsilon (ε) is a ratio and varies between 0 and 1. For most natural materials, it ranges between 0.7 and 0.95.

Table: Emissivity of Common Materials.

Material	Emissivity
Clear water	0.98-0.99
Wet snow	0.98-0.99
Human skin	0.97-0.99
Rough ice	0.97-0.98
Vegetation	0.96-0.99
Wet soil	0.95-0.98
Asphalt concrete	0.94-0.97

Brick	0.93-0.94
Wood	0.93-0.94
Basalt rock	0.92-0.96
Dry mineral soil	0.92-0.94
Paint	0.90-0.96
Dry vegetation	0.88-0.94
Dry snow	0.85-0.90

Kinetic temperature is the surface temperature of a body/ground and is a measure of the amount of heat energy contained in it. It is measured in different units, such as in Kelvin (K); degrees Centigrade (°C); degrees Fahrenheit (°F).

The radiant temperature calculated from the radiant energy emitted is in most cases smaller than the true, kinetic temperature (T_{kin}) that could be measure with a contact thermometer on the ground. The reason is that most objects have a Kinetic temperature emissivity lower than 1.0 and radiate incompletely. To calculate the true T_{kin} from the T_{rad}, we need to know or estimate the emissivity. The relationship between T_{kin} and T_{rad} is:

$$T_{rad} = \varepsilon^{1/4} T_{kin}$$

Concept of Black body

An object radiates unique spectral radiant flux depending on the temperature and emissivity of the object. This radiation is called thermal radiation because it mainly depends on temperature and it can be expressed in terms of 'black body' theory. Black body is a theoretical object that absorbs and then emits all incident energy at all wavelengths. This means that the emissivity of such an object is by definition. However, true black-bodies do not exist in nature, although some materials (eg, clean, deep water radiating between 8 to 12 μm) come very close.

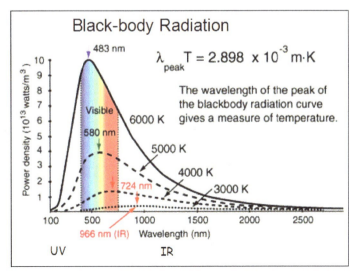

Materials that absorb and radiate only a certain fraction compared to a blackbody are called 'grey bodies'. The fraction is a constant for all wavelengths. Hence, a grey-body curve is identical in shape to a black-body curve, but the absolute values are lower as it does not radiate as perfectly as a black-body.

A third group is the 'selective radiators'. They also radiate only a certain fraction of a black-body, but this fraction changes with wavelength. A selective radiator may radiate perfectly in some wavelengths, while acting as a very poor radiator in other wavelengths. The radiant emittance curve of a selective radiator can then also look quite different from an ideal, black-body curve.

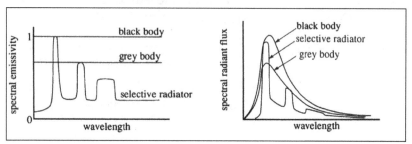

Spectral emissivities and radiant exitances for a black body, a gray body and selective radiator.

The fraction of energy that is radiated by a material compared to a true blackbody is also referred to as emissivity ($\varepsilon\lambda$). Hence, emissivity is defined as:

$$\text{Emissivity} = \frac{\text{Radiant energy of an object}}{\text{Radiant energy of a black body with the same temperature as the object}}$$

Or,

$$\varepsilon\lambda = \frac{M_{T,\lambda}}{M_{\lambda,T}^{BB}}$$

Where, $M_{T,\lambda}$ is the radiant emittance of a real material at a given temperature, $M_{\lambda,T}^{BB}$ is a radiant emittance of a black-body at the same temperature.

- A black body has $\varepsilon\lambda = 1$.
- A gray body has $\varepsilon\lambda = $ constant.
- A selective radiator has $\varepsilon\lambda = \text{fn}(\lambda)$.

Data Acquisition: Modes and Platforms

There are three different aspects which have to be taken into account while mode of thermal data acquisition is considered:

- Active versus passive mode: Most of the thermal sensors acquire data passively, i.e. they measure the radiations emitted naturally by the target/ground. Data can also be acquired in the TIR actively deploying laser beams (LIDAR). However, these techniques are not well researched and are only in the infancy.

- Broad band versus multispectral mode: For the broad band thermal sensing, in general the 8 to 14 µm atmospheric window is utilised. However, some space borne thermal sensors such as Landsat Thematic Mapper Band 6 operates in the wavelength range of 10.4 to 12.6 µm to avoid the ozone absorption peak which is located at 9.6 µm. The multispectral thermal channels, such as those in the ASTER platform, are targeted specially for geological applications.

- Daytime versus night-time acquisition: Thermal data can be acquired during day and night also. For some applications it is useful to have data from both the times. However, for many applications night-time or more specifically pre-dawn images are preferred as during this time the effect of differential solar heating is the minimal. The platforms for such data acquisitions range from satellites, aircrafts to ground based scanners.

Types of Scanners used in Thermal Remote Sensing

Across-track Thermal Scanners

For Across-track thermal scanners, Daedalus DS-1260, DS-1268, and Airborne Multispectral Scanner are used as thermal Infrared Multispectral Scanners. These scanners provide most of the useful high spatial and spectral resolution thermal infrared data for monitoring the environment. The DS-1260 records data in 10 bands including a thermal-infrared channel (8.5 to 13.5 µm). The DS-1268 incorporates the thematic mapper middle-infrared bands (1.55 - 1.75 µm and 2.08 - 2.35 µm). The AMS contains a hot-target, thermal-infrared detector (3.0 to 5.5 µm) in addition to the standard thermal-infrared detector (8.5 to 12.5 µm).

The detectors are cooled to low temperatures (-196 °C, -243 °C, 73 °K) using liquid helium or liquid nitrogen. Cooling the detectors insures that the radiant energy (photons) recorded by the detectors comes from the terrain and not from the ambient temperature of objects within the scanner itself.

Push-broom Linear and Area Array Charge-coupled Device (CCD) Detectors

For this, solid-state microelectronic detectors are used. Those are smaller in size (e.g. 20 × 20 mm) and weight. They require less power to operate, have fewer moving parts, and are more reliable. Each detector in the array can view the ground resolution element for a longer time, allowing more photons of energy from within the IFOV to be recorded by the individual detector. It helps in improving radiometric resolution (the ability to resolve smaller temperature differences). Each detector element in the linear or area array is fixed relative to all other elements, therefore, the geometry of the thermal infrared image is much improved relative to that produced by an across-track scanning system and some linear and area thermal detectors do not even require the cooling apparatus.

Forward-looking Infrared (FLIR) Systems

Forward-Looking Infrared type systems are calibrated with aircraft that look obliquely ahead of the aircraft and acquire high-quality thermal infrared imagery, especially at night. FLIR systems collect the infrared energy based on the same principles as an across-track scanner, except that the mirror points forward about 45° and projects terrain energy during a single sweep of the mirror onto a linear array of thermal infrared detectors.

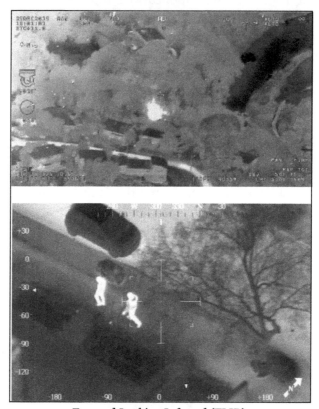

Forward Looking Infrared (FLIR).

Thermal Sensors and Satellites

Thermal sensors or scanners detect emitted radiant energy. Due to atmospheric effects these sensors usually operate in the 3 to 5 μm or 8 to 14μm range. Most thermal remote sensing of Earth features is focused in the 8 to 14 μm range because peak emission (based on Wien's Law) for objects around 300K (27 °C or 80 °F) occurs at 9.7μm. Many thermal imaging sensors are on satellite platforms, although they can also be located on-board aircraft or on ground-based systems. Many thermal systems are multispectral, meaning they collect data on emitted radiation across a variety of wavelengths.

Thermal Sensors

Thermal Infrared Multispectral Scanner (TIMS)

NASA and the Jet Propulsion Laboratory developed the Thermal Infrared Multispectral Scanner (TIMS) for exploiting mineral signature information. TIMS is a multispectral scanning system with six different bands ranging from 8.2 to 12.2 μm and a spatial resolution of 18m. TIMS is mounted on an aircraft and was primarily designed as an airborne geologic remote sensing tool. TIMS acquires mineral signature data that permits the discrimination of silicate, carbonate and hydrothermally altered rocks. TIMS data have been used extensively in volcanology research in the western United States, Hawaiian islands and Europe. The multispectral data allows for generate of three-band color composites similar other multispectral data. Many materials have varying emissivities and can be identified by the variation in emitted energy.

The thermal image to the right was captured the Thermal Infrared Multispectral Scanner (TIMS) and is a thermal image of Death Valley California. A color composite has been produced using three of the thermal bands collected by TIMS. There are a variety of different materials and minerals in Death Valley with varying emissivities. In this image Thermal Band 1 (8.2 - 8.6 μm) is displayed in blue, Thermal Band 3 (9.0 - 9.4 μm) is displayed in green and Thermal Band 5 (10.2 - 11.2 μm) is displayed in red. Alluvial fans appear in shades of reds, lavender, and blue-greens; saline soils in yellow; and different saline deposits in blues and greens.

Advanced Spaceborne Thermal Emission and Reflection Radiometer (ASTER)

Advanced Spaceborne Thermal Emission and Reflection Radiometer (ASTER) is a sensor on-board the Terra satellite. In addition to collecting reflective data in the visible, bear and shortwave infrared, ASTER also collects thermal infrared data. ASTER has five thermal bands ranging from 8.1 to 11.6 μm with 90m spatial resolution. ASTER data are used to create detailed maps of surface temperature of land, emissivity, reflectance, and elevation.

Moderate-resolution Imaging Spectroradiometer (MODIS)

MODIS has a high spectral resolution and collects data in a variety of wavelength. Similar to ASTER, MODIS collects reflective data and emitted, thermal data. MODIS has several bands that collect thermal data with 1000m spatial resolution. MODIS has high temporal resolution with a one to two day return time. This makes it an excellent resource for detecting and monitoring wildfires. One of the products generated from MODIS data is the Thermal Anomalies/Fire product which detects hotspots and fires.

MODIS Thermal Anomalies/Fire data.

Landsat

A variety of the Landsat satellites have carried thermal sensors. The first Landsat satellite to collect thermal data was Landsat 3, however this part of the sensor failed shortly after the satellite was launched. Landsat 4 and 5 included a single thermal band (band 6) on the Thematic Mapper (TM) sensor with 120m spatial resolution that has been resampled to 30m. A similar band was included on the Enhanced Thematic Mapper Plus (ETM+) on Landsat 7. Landsat 8 includes a separate thermal sensor known the Thermal Infrared Sensor (TIRS). TIRS has two thermal bands, Band 10 (10.60 - 11.19µm) and Band 11 (11.50 - 12.51µm). The TIRS bands are acquired at 100 m spatial resolution, but are resampled to 30m in the delivered data products.

Landsat TIRS and Applications

Irrigation accounts for 80% of fresh water use in the U.S and water usage has become an increasingly important issue, particularly in the West. Thermal infrared data from Landsat 8 is being used to estimate water use. Landsat 8 data, including visible, near infrared, mid-infrared, and thermal data are fed into a relatively sophisticated energy balance model that produces evapotranspiration maps. Evapotranspiration (ET) refers to the conversion of water into water vapor by the dual process of evaporation from the soil and transpiration (the escape of water though plant's stomata). For vegetated land, ET is synonymous with water consumption. Landsat data enable water resources managers and administrators to determine how much water was consumed from individual fields.

Conducting IR Surveys

- Time of day: Thermal infrared scanners are used most frequently at night when there is no interference from reflected solar radiation. The usual flying time is just before dawn, when the effects of differential solar heating are at their lowest level. Night time imagery is necessary for most geologic applications because the thermal effects of differential solar heating and shadowing are greatly reduced. On daytime images topography is typically the dominant expression because of these differential solar effects. As the radiant temperatures are relatively constant in the pre-dawn hours, thermal imagery obtained during such hours is preferable for interpretation purposes.

- Spatial resolution: Spatial resolution of IR imagery depends upon the flight altitude and the instantaneous field of view of the detector. At an altitude of 2000m with an instantaneous field of view of 3 milliradians, the ground resolution becomes 6m. Flight altitude of 2000 m. is considered best for conducting IR surveys. IFOV typically ranges from 2-3 milliradians.

- Wavelength Bands: Thermal IR images may be acquired at wavelength bands of 3-5mm band corresponds to the radiant energy peak for temperatures of 600 °K and greater that are associated with lava flows, fires and other hot features. The 8-14μm band spans the radiant energy peak for a temperature of 300 °K. It is the ambient temperature of the earth having the radiant energy peak at 9.7μm. Hence 8-14μm images are optimum for terrain mapping whereas 3-5μm images are optimum for mapping hot targets such as fires.

- Orientation and altitude of flight lines: Flight altitude influences image scale, lateral ground coverage and spatial resolution. For geologic projects, it is useful to know the regional structural strike or tectonic grain of the area in advance of an IR survey. It helps in determining the optimum orientation of flight lines. If flight lines are oriented normal to the regional strike, it may mask linear geologic features. Hence it is preferable to orient flight lines with or at an acute angle to the regional strike.

- Ground measurements: Weather and surface conditions play a large role in determining terrain expression on IR images. It may be useful to collect ground information on weather conditions, soil moisture and vegetation at the time of IR survey. Ground measurements are most valuable if they are made at localities that have image signatures.

Advantages of Thermal Imagery

Applications of Thermal infrared remote sensing can be broadly classified into two categories one in which surface temperature is governed by man-made sources of heat and other in which it is governed by solar radiation. In the former case the technique has been used from airborne platforms for determining heat losses from buildings and other engineering structures. In the latter case, TIR remote sensing has been used for identifying crop types, soil moisture, measuring water stress, etc. Some of the applications areas of TIR remote sensing are given below:

- Geology: Denser rocks such as basalt and sandstone have higher thermal inertia and on night thermal IR images they show warm signature compared to less dense rocks such as siltstone, cinders etc. Hence differentiation of rock types is possible. Faults may be marked by cooler linear anomalies caused by evaporative cooling of moisture trapped along the fault zone. Folds may be indicated by thermal patterns caused by out-crops of different rock types. Surface temperatures of volcanic terrain may be mapped.

- Military: Military applications of thermal mapping are varied and often classified. Unusual concentration of troops, weaponry, military vehicles, jungle trails etc. can be identified by the tonal changes in the thermal imagery.

- Hydrology: Cool underground springs that discharge into warmer courses of water may be discovered by thermal imagery. Hot industrial effluents that pollute water may be detected and the diffusion pattern may be mapped. Hot springs are clearly identifiable in IR imagery.

- Agriculture: Thermal images have been utilized for the identification of crop species and soil types, for detecting crop diseases, for making animal censuses and for determining relative moisture content of various soils.

- Botany: Leaf temperatures can be remotely measured using 3 to 5µm band. Such knowledge is useful for assessing plant health, age, relative water supply or degree of irrigation. Thermal damage of frost damage to fruit groves can also be assessed.

- Forestry: Forest fires beneath a forest cover story can be located with thermal IR. During extensive fires, damage caused can be assessed even under excessive smoke conditions.

- Heat Loss Surveys: To survey heated buildings, factories and buried steam lines for anomalous hot spots that may indicate leakage and poorly insulated roofs.

Hyperspectral Remote Sensing

The 'hyperspectral remote sensing' is developed in mid-80 and considered to be the most significant recent break-through. Since then it has been widely used in the detection and identification of minerals, vegetation, artificial materials and soil background. Hyperspectral remote sensing is technologically more developed than multispectral remote sensing and its sensors have the ability to acquire images in many narrow spectral bands that are found in the electromagnetic spectrum from visible, near infrared, medium infrared to thermal infrared.

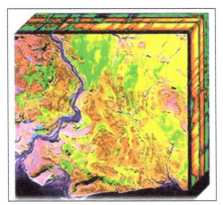

Two dimensional projection of a hyperspectral cube.

Hyperspectral sensors capture energy in 200 bands or more which means that they continuously cover the reflecting spectrum for each pixel in the scene. Bands characteristic for these types of sensors are continuous and narrow (10-20 nm), allowing an in depth examination of features and details on Earth. Hyperspectral sensors are working in hundreds of bands, but not the number of bands defines the sensor as being hyperspectral. The criteria underlying the classification of sensors as hyperspectral are bandwidth and the continuous nature of the records. For example, a sensor that only works in 20 bands may be considered hyperspectral if all these bands are adjacent and with a 10 nm width.

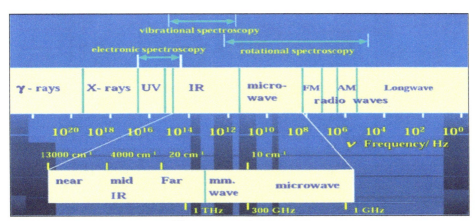

Electromagnetic spectrum.

Hyperspectral images provide ample spectral information to identify and distinguish spectrally unique materials through extracting the information upto sub-pixel scale. In this way hyperspectral, imagery provides the potential for more accurate and detailed information extraction than possible with any other type of remotely sensed data. Hyperspectral records are based on spectroscopy in the range of 0.40–2.50 μm where hyperspectral sensors are working. Field and laboratory spectrometers usually measure reflectance at many narrow, closely spaced wavelength bands, so that the resulting spectra appear to be continuous curves. When a spectrometer is used in an imaging sensor, the resulting images record a reflectance spectrum for each pixel in the image. The identification of a target material is determined by comparison of its spectral reflectance curve with 'library spectra' of known materials measured in the field or in the laboratory.

The Imaging Spectrometer

Hyperspectral images are produced by instruments called imaging spectrometers. The development of these complex sensors has involved the convergence of two related but distinct technologies: spectroscopy and the remote imaging of Earth and planetary surfaces. Spectroscopy is the study of light that is emitted by or reflected from materials and its variation in energy with wavelength. Spectroscopy deals with the spectrum of sunlight that is diffusely reflected (scattered) by materials at the Earth's surface. Instruments called spectrometers (or spectro radiometers) are used to make ground-based or laboratory measurements of the light reflected from a test material. An optical dispersing element such as a grating or prism in the spectrometer splits this light into many narrow, adjacent wavelength bands and the energy in each band is measured by a separate detector. By using hundreds or even thousands of detectors, spectrometers can make spectral measurements of bands as narrow as 0.01 micrometers over a wide wavelength range, typically at least 0.4 to 2.4 micrometers (visible through middle infrared wavelength ranges).

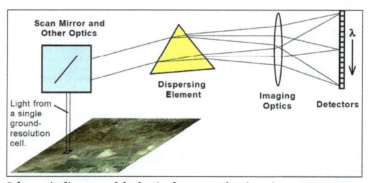

Schematic diagram of the basic elements of an imaging spectrometer.

Remote imagers are designed to focus and measure the light reflected from many adjacent areas on the Earth's surface. In many digital images, sequential measurements of small areas are made in a consistent geometric pattern as the sensor platform moves and subsequent processing is required to assemble them into an image. Until recently, imagers were restricted to one or a few relatively broad wavelength bands by limitations of detector designs and the requirements of data storage, transmission, and processing. Recent advances in these areas have allowed the design of imagers that have spectral ranges and resolutions comparable to ground-based spectrometers.

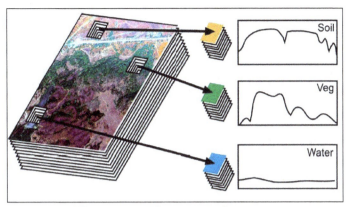

The concept of hyperspectral imagery.

Spectral Reflectance

In reflected-light spectroscopy, the fundamental property that we want to obtain is spectral reflectance that is the ratio of reflected energy to incident energy as a function of wavelength. Reflectance varies with wavelength for most materials because energy at certain wavelengths is scattered or absorbed to different degrees. A reflectance curve can be prepared by plotting of reflectance and wavelength on x and y-axis. The overall shape of a spectral curve and the position and strength of absorption bands in many cases can be used to identify and discriminate different materials. For example, vegetation has higher reflectance in the near infrared range and lower reflectance of red light than soils.

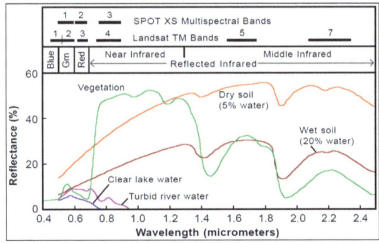

Representative spectral reflectance curves for several common Earth surface materials over the visible light to reflected infrared spectral range.

Mineral Spectra

Every material is formed by chemical bonds, and has the potential for detection with spectroscopy. Spectroscopy can be used to detect individual absorption features due to specific chemical bonds in a solid, liquid, or gas. Solids can be either crystalline (i.e. minerals) or amorphous (like glasses). In inorganic materials such as minerals, chemical composition and crystalline structure control the shape of the spectral curve and the presence and positions of specific absorption bands. Wavelength specific absorption may be caused by the presence of particular chemical elements or ions, the ionic charge of certain elements, and the geometry of chemical bonds between elements, which is governed in part by the crystal structure.

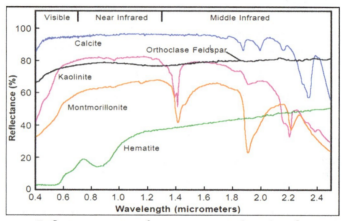

Reflectance spectra of some representative minerals.

In the spectrum of hematite (an iron-oxide mineral), the strong absorption in the visible light range is caused by ferric iron (Fe^{+3}). In calcite, the major component of limestone, the carbonate ion (CO_3^{-2}) is responsible for the series of absorption bands between 1.8 and 2.4 micrometers (µm). Kaolinite and montmorillonite are clay minerals that are common in soils. The strong absorption band near 1.4µm in both spectra, along with the weak 1.9µm band in kaolinite, are due to hydroxide ions (OH^{-1}), while the stronger 1.9µm band in montmorillonite is caused by bound water molecules in this hydrous clay. In contrast to these examples, orthoclase feldspar, a dominant mineral in granite, shows almost no significant absorption features in the visible to middle infrared spectral range.

Plant Spectra

Reflectance spectra of green vegetation differ if compared to a spectral curve for dry or yellowed leaves. Different portions of the spectral curves for vegetation are shaped by different plant components. The spectral reflectance curves of healthy green plants have a characteristic shape that is dictated by various plant attributes. In the visible portion of the spectrum, the curve shape is governed by absorption effects from chlorophyll and other leaf pigments. Chlorophyll absorbs visible light very effectively but absorbs blue and red wavelengths more strongly than green, producing a characteristic small reflectance peak within the green wavelength range. As a consequence, healthy plants appear to us as green in color. Reflectance rises sharply across the boundary between red and near infrared wavelengths to values of around 40 to 50% for most plants.

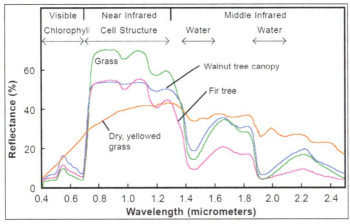

Reflectance spectra of different types of green vegetation compared to a spectral curve for senescent (dry, yellowed) leaves.

This high near-infrared reflectance is primarily due to interactions with the internal cellular structure of leaves. Most of the remaining energy is transmitted, and can interact with other leaves lower in the canopy. Leaf structure varies significantly between plant species, and can also change as a result of plant stress. Thus, species type, plant stress, and canopy state all can affect near infrared reflectance measurements. Beyond 1.3μm, reflectance decreases with increasing wavelength, except for two pronounced water absorption bands near 1.4 and 1.9μm. At the end of the growing season leaves, lose water and chlorophyll. Therefore, near infrared reflectance decreases and red reflectance increases, creating the familiar yellow, brown, and red leaf colors of autumn.

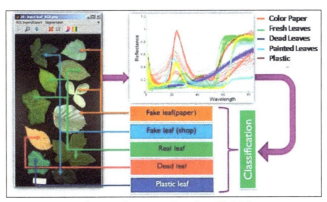

Spectral curve for natural and artificial leaves.

Spectral Libraries

Several libraries of reflectance spectra of natural and man-made materials are available for public use. These libraries provide a source of reference spectra that can aid the interpretation of hyperspectral images.

- ASTER Spectral Library: This library has been made available by NASA as part of the Advanced Spaceborne Thermal Emission and Reflection Radiometer

(ASTER) imaging instrument program. The ASTER spectral library currently contains nearly 2000 spectra, including minerals, rocks, soils, man-made materials, water, and snow. Many of the spectra cover the entire wavelength region from 0.4 to 14µm.

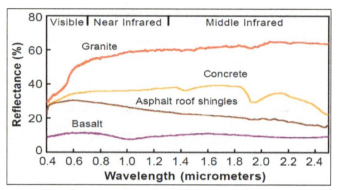

Sample spectra from the ASTER Spectral Library.

- USGS Spectral Library: The United States Geological Survey Spectroscopy Lab has compiled a library of about 500 reflectance spectra of minerals and a few plants over the wavelength range from 0.2 to 3.0µm.

Hyperspectral Sensors

In hyperspectral remote sensing there are two types of sensors:

- Spectral Sensors on Satellites.
- Spectral Sensors on Aircraft.

The Hyperion EO-1 sensor was launched in November 2000 by NASA with the purpose of taking hyperspectral images from space in order to create mineralogical mapping. It works in the spectral range 0.40–2.50µm with 242 bands. It has spectral resolution of about 10 nm and spatial resolution of 30 meters. The data is taken from an altitude of 705 km. Hyperion is a push-broom instrument that takes pictures with a radiometric resolution of 8 bits with a band width of 7.5 km and being perpendicular on the movement of the satellite. The system used for taking images is formed of two spectrometers: (i) one working in the visible/near infrared (VNIR) (0.4–1.0µm) and (ii) another one in shortwave infrared (SWIR) (0.9–2.5µm).

Airborne Visible Sensor/Infrared Imaging Spectrometer (AVIRIS) developed by NASA/Jet Propulsion Laboratory (JPL) is a new in terms of hyperspectral systems attached to planes. AVIRIS was started in 1998 and its sensor is mounted on a Twin Otter aircraft flying at low altitude, taking pictures with a spatial resolution ranging between 2 and 4 meters. AVIRIS sensor can also take images from an altitude of 20 km with a spatial resolution of 20 meters, from a bandwidth of 10.5 kilometers. It is working in bands of 224, with spectral range from 0.40 to 2.50µm. The sensor is a Whiskbroom system that uses a scanning system for acquiring data on the transverse direction of advancement. Most hyperspectral sensors are mounted on aerial platforms and less on the satellite. However, sensors on satellites have the capacity to provide global coverage at regular intervals.

Table: Characterization of Hyperspectral Sensors.

Types of sensor	Manufacturer	Number of Bands	Spectral Range (µm)
Hyperspectral Sensors on Satellite			
FTHSI on MightySat II	Air Force Research Lab	256	0.35 to 1.05
Hyperion on EO-1	NASA Goddard Space Flight Center	220	0.4 to 2.5
Hyperspectral Sensors on Aircraft			
AVIRIS (Airborne Visible Infrared Imaging Spectrometer)	NASA, USA	224	0.4 to 2.5
HYDICE (Hyperspectral Digital Imagery Collection Experiment)	Naval Research Lab	210	0.4 to 2.5
PROBE-1	Earth Search Sciences Inc., USA	128	0.4 to 2.5
CASI (Compact Airborne Spectrographic Imager)	ITRES Research Limited, Canada	up to 228	0.4 to 1.0
HyMap	Integrated Spectronics Pt. Ltd., Australia	100 to 200	Visible to thermal infrared
EPS-H (Environmental Protection System)	GER Corporation	VIS/NIR (76), SWIR1 (32), SWIR2 (32), TIR (12)	VIS/NIR (0.43-1.05), SWIR1 (1.5-1.8), SWIR2 (2.0-2.5), TIR (8-12.5)
AIS 7915 (Digital Airborne Imaging Spectrometer)	GER Corporation	VIS/NIR (32), SWIR1 (8), SWIR2 (32), MIR (1), TIR (6)	VIS/NIR (0.43-1.05), SWIR1 (1.5-1.8), SWIR2 (2.0-2.5), MIR (3.0-5.0), TIR (8.7-12.3)
DAIS 21115 (Digital Airborne Imaging Spectrometer)	GER Corp., USA	VIS/NIR (76), SWIR1 (64), SWIR2 (64), MIR (1), TIR (6)	VIS/NIR (0.40 to 1.0), SWIR1 (1.0 to 1.8), SWIR2 (2.0 to 2.5), MIR (3.0 to 5.0), and TIR (8.0 to 12.0)
AISA (Airborne Imaging Spectrometer)	Spectral Imaging Ltd., Finland	up to 288	0.43 to 1.0

Image Analysis

The hyperspectral images provide the fine spectral resolution needed to characterize the spectral properties of surface materials but the volume of data in a single scene is very vast. The difference in spectral information between two adjacent wavelength bands is typically very small and their grayscale images therefore appear nearly identical.

Finding appropriate tools and approaches for visualizing and analyzing the essential information in a hyperspectral scene needs expertise and active research. Followings are some of the techniques used for analyzing the spectral content of hyperspectral images:

- Match Each Image Spectrum: One approach to analyzing a hyperspectral image is to match each image spectrum individually to the reference reflectance spectra in a spectral library. This approach requires an accurate conversion of image spectra to reflectance.

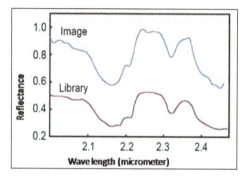 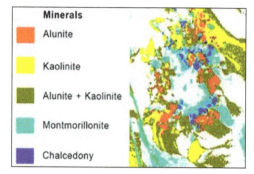

Sample image spectrum and a matched spectrum of the mineral alunite from the USGS Spectral Library.

Mineral map for part of the Cuprite AVIRIS scene, created by matching image spectra to mineral spectra in the USGS Spectral Library.

- Spectral Matching Methods: The shape of a reflectance spectrum can usually be broken down into two components: (i) broad, smoothly changing regions that define the general shape of the spectrum and (ii) narrow, trough-like absorption features. This distinction leads to two different approaches to matching image spectra with reference spectra. One common matching strategy is to match only the absorption features in each candidate reference spectrum and ignores other parts of the spectrum. Many other materials, such as rocks and soils, may lack distinctive absorption features then these spectra are characterized by their overall shape.

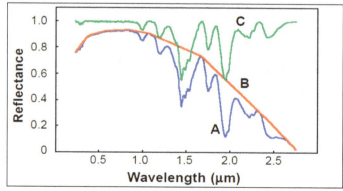

Reference spectrum for minerals gypsum (A) with several absorption features. Curve B shows the continuum for the spectrum, and C the spectrum after removal of continuum.

- Linear Unmixing: Linear unmixing is an alternative approach to simple spectral matching. The set of spectrally unique surface materials existing within a scene are referred to as the spectral endmembers for that scene. Linear Spectral Unmixing exploits the theory that the reflectance spectrum of any pixel is the result of linear combinations of the spectra of all endmembers inside that pixel. If the spectra of all endmembers in the scene are known, then their abundances within each pixel can be calculated from each pixel's spectrum. The results of Linear Spectral Unmixing include one abundance image for each endmember. The pixel values in these images indicate the percentage of the pixel made up of that endmember. For example, if a pixel in an abundance image for the endmember quartz has a value of 0.90, then 90% of the area of the pixel contains quartz. An error image is also usually calculated to help evaluate the success of the unmixing analysis.

  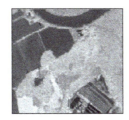 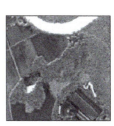

AVIRIS scene   Soil fraction   Vegetation fraction   Water fraction

References

- Image-interpretation-of-remote-sensing-data: geospatialworld.net, Retrieved 25, July 2020
- Microwave-engineering-microwave-devices: tutorialspoint.com, Retrieved 10, April 2020
- Geo-Thermal: lkouniv.ac.in, Retrieved 05, June 2020

Chapter 2
Remote Sensing and Image Sensing

Photogrammetry involves the techniques used to obtain information of real world objects through recording, measuring and interpreting photographic images. Taking photographs from an aircraft or other flying object is known as aerial photography. This chapter discusses in detail all aspects related to remote sensing and image processing like photogeology, image rectifications, image enhancement etc.

Aerial Photography and Photogrammetry

Aerial Photography is defined as art, science and technology of taking aerial photographs from an air-borne platform. Probably Gasper Felix Tournachon "Nadar" took very first aerial photograph in 1858 of a village of Petit Bicetre (France) from a balloon. During World War-I aerial photography got major momentum of development. Air photos were taken for reconnaissance from fighter planes and pigeons. Small lightweight cameras were attached to the birds and a timer was set to take pictures every 30 seconds as it flew.

Now, aerial photographs are taken from aircraft to capture series of images using a large roll of special photographic film. The film is processed and cut into negatives. The common sizes of negatives are 23 × 23 cm. The basis of aerial photography is light sensitive chemicals in the film emulsion. These chemicals may react to ultraviolet, visible and near infra-red portions of the spectrum from 0.3 μm to 0.9 μm wavelength. To plan an aerial photographic mission we must define final product then determine camera system, other material required, determine flight pattern and setting shutter timing for end lap and overlap.

Photogrammetry

Photogrammetry can be defined as the science and art of determining qualitative and quantitative characteristics of objects from the images recorded on photographic emulsions without coming in physical contact with the objects. Here information is obtained through processes of recording patterns of electromagnetic radiant energy, predominantly in the form of photographic images. Objects are identified and qualitatively described by observing photographic image characteristics such as shape, pattern, tone, and texture. Photogrammetry also allows for the extraction of three-dimensional features from remotely sensed data.

Types of Aerial Photograph

Vertical Aerial Photograph

Vertical aerial photography is an aerial photography technique where the shots are taken from directly above the subject of the image. Allowable tolerance is usually + 3° from the perpendicular (plumb) line to the camera axis. This method of aerial photography is also referred as "overhead aerial photography." In vertical aerial photograph, the lens axis is perpendicular to the surface of the earth. In vertical photograph, we may see flat and map-like image of the rooftops and canopies of the building and structure being photographed.

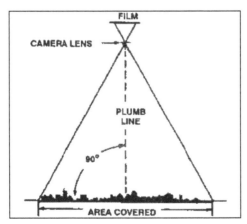

There are three common ways that vertical aerial photography can be conducted:

- Low Altitude: For this particular shot, the resulting images will show bigger and closer shots of the subject and its surroundings.

- Medium Altitude: Here, the resulting images of the subject and the surroundings are smaller than those produced in low altitude vertical aerial photography.

- High Altitude: The images of the subject and its surroundings produced from high altitude vertical aerial photography are way smaller than those produced from low altitude and medium altitude vertical aerial photography. Nonetheless, they are able to cover a wider section of the land.

Oblique Photography

The word oblique means having a sloping direction or angular position. Therefore, Photographs taken at an angle are called oblique photographs. Oblique Photography is of two types:

- Low Oblique Aerial Photography: Low oblique aerial photograph is a photograph taken with the camera inclined about 30° from the vertical. In this type of photograph horizon is not visible. The ground area covered is a trapezoid, although the photo is square or rectangular. No scale is applicable to the entire photograph, and distance cannot be measured. Parallel lines on the ground are not parallel on this photograph; therefore, direction (azimuth) cannot be measured. Relief is detectable but distorted.

- High Oblique Aerial Photography: The high oblique is a photograph taken with the camera inclined about 60° from the vertical. In this type of aerial photograph horizon is visible. It covers a very large area. The ground area covered is a trapezoid, but the photograph is square or rectangular. Distances and directions are not measured on this photograph for the same reasons that they are not measured on the low oblique. Relief may be quite detectable but distorted as in any oblique view.

Basics of Aerial Camera

An aerial camera is a highly specialized camera, designed for use in aircraft and containing a mechanism to expose the film in continuous sequence at a steady rate. They are referred to as "passive sensors" because they detect and capture the natural light reflected from objects.

An aerial camera is a mechanical optical instrument with automatic and electronic elements. It is designed for obtaining aerial photographs of the earth's surface from an airplane or other type of aircraft. Aero-camera differs from ordinary camera and has specific features like fully automatic operation, shock absorbing support frame, large picture format, and rapid frame advance. Apart from these, aero-camera is accomplished with photographing from great distance, rapid movement and vibration during exposure. The world's first aerial camera for area photography from an areophane was invented by the Russian army engineer V. F. Potte during World War I.

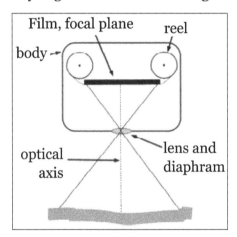

Aerial cameras may have one or more lenses for plan views, perspectives and panoramic survey. Basic features of aerial cameras are their focal length, negative size, and minimum exposure time, which is as short as 1/1000 sec in Soviet aerial cameras. Standard Soviet aerial cameras designed for topographical surveys, are having 18 × 18 cm negative size, have focal lengths from 50 to 500 mm and corresponding field-of-view angles from 150° to 30°. Aerial cameras of either type are used for black and white or color aerial photographic surveying.

Aerial cameras with varying focal lengths, beginning with 88 mm, are also used aboard. With the most popular 23 × 23 cm negative size, this corresponds to field-of view angles up to 125°. Like other cameras, aerial camera also has basic features of Lens, Shutter and Diaphragm, working as focal plane/focal length, controlling exposure speed and aperture respectively.

Focal Plane and Focal Length

Focal plane is the flat surface where film is held. Focal length is the distance from the focal plane to approximately the center of the camera lens. Thin lens equation is:

$$\frac{1}{f} = \frac{1}{o} + \frac{1}{i}$$

f = focal length of the camera, o = distance between object and camera and i = distance between lens and image plane.

Most metric aerial cameras have a fixed focal length such as 152 mm and 305mm. Military photoreconnaissance operations commonly employ lenses 3 to 6 feet to obtain detailed photographs from extremely high altitudes.

Types of Aerial Cameras

- Single-lens mapping (metric) cameras: It provides highest geometric and radiometric quality of aerial photography to map the planimetric (x, y) location of features and to derive topographic contour map. Individual exposures are typically 23 × 23 cm.

- Multiple-lens cameras: Each of the lenses (camera) simultaneously records photographs of the same area, but using different film and filter combination. This creates multiple band photographs.

- Panoramic camera: This type of camera uses a rotating lens (or prism) to produce a narrow strip of imagery perpendicular to the flight line. It is commonly used by military but much less in civilian applications due to poor geometric integrity.

- Digital camera: Digital camera uses "charge-coupled-device (CCD) detectors. These detectors are arranged in a matrix and located at the film plane. A digital camera takes light and focuses it via the lens onto a sensor made out of silicon. It is made up of a grid of tiny photosites that are sensitive to light. Each photosite is usually called a pixel ("picture element").

There are millions of individual pixels in the sensor of a digital camera. Advantage with digital camera is that it records and store photographic images in digital form. These images can be stored directly in the camera or can be uploaded onto a computer or printer later on. To replicate the spatial resolution of standard 9x9 inches metric aerial photograph, a digital camera would require approximately 20000 × 20000 detectors.

Aerial Photograph Filtration

Lens filters are transparent or translucent glass or gelatin elements that attach to the front of a camera lens. They protect the camera lens, alter the characteristics of light passing through the lens or add special effects and colour to an image. Photographic filters are used to achieve image enhancement effects that can change the tone and mood of the photographs. Filters work on the theories of additive colour and subtractive colour.

- Additive colour: Blue, green and red are considered to be additive colour and also known as primary colour. These primary colours can be mixed to create all colour shades.

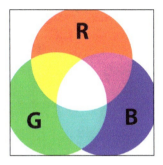

- Subtractive colour: Yellow, cyan and magenta are the subtractive colour. These colours are also known as secondary colour. A subtractive colour model explains the mixing of a limited set of dyes, inks, paint pigments or natural colourants to create a wider range of colours, each the result of partially or completely subtracting some wavelengths of light. Filters filter out certain types of unwanted wavelengths of light before they can reach the film plane and expose the film. A filter will appear the colour of light that is allowed to pass through.

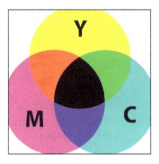

- Haze filter: When collecting natural colour aerial photograph, it is desired to eliminate much of the scattering of ultraviolet radiation caused by atmosphere haze. Haze filters were developed to absorb light shorter than 400nm.

Without Filter   With Filter

- Yellow filter (minus blue filter): When collecting colour infrared aerial photography, yellow filter is used, which subtract almost all of the blue light (wavelength < 500nm). This reduces the effect of atmospheric Rayleigh scattering. It absorbs blue and allows green and red light to be transmitted. A mixture of red and green is yellow.

Yellow Filter.

- Band pass filter: Band pass filter configure a film and filter combination so that the camera only records a very specific band of reflected EM energy.

Band Pass Filter.

- Polarization filter: A Polarizing filter reduces atmospheric haze, but also reduces reflected sunlight. The most typical function of a Polarizer is to remove reflections from water and glass. The resulting image is free of reflected light, and transparent objects like glass are free of reflections. It allows the vibration of a light ray in just one plane to be passed.

Types of Film

- Panchromatic: Panchromatic film is also called as 'black and white' film. It is sensitive to the same range of light wavelength as perceived by the human eye. Panchromatic film is most commonly used for planimetric and topographic map. A yellow filter is normally used for exposure on panchromatic film to reduce the fogging effect caused by atmospheric haze. Unfiltered panchromatic film is used for penetration through clear water.

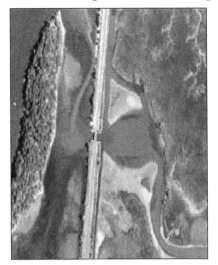 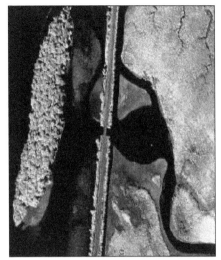

Panchromatic.　　　　　　　　Black and White Infrared.

- Black and White Infrared (IR): It is sensitive to a range of wavelength that includes the green, red and near infrared position of the spectrum. It has great ability to differentiate different types of vegetation. Healthy deciduous vegetation is recorded in light tones and coniferous registers in dark tones. It is also useful for differentiating dry and moist soils. In Black IR film moist soil appears in dark tone and dry soil in light tones. The NIR wavelength (0.7-1.0) cannot be perceived by the human eye, so they provide information that beyond the human perception system.

- Natural colour: It is often called true colour is sensitive to the same wavelength of light as perceived by the human eye. It is especially useful for identifying soil types, rock types and surficial deposits, water surface patterns and various forms of polluted water. It has good penetration qualities and is therefore valuable for recording underwater features. Its penetration through clear water can exceed 25 m. Colour photography is also useful for detecting forest damage caused by various insects.

- Colour Infrared (CIR): It was developed during World War II for detecting camouflaged military targets. It is also called as false colour (or FCC). Like natural colour, they are usually displayed using the RGB colour system to re-create the same colour as on the photo print. Vegetation usually appears red on these images, thus the term false colour. CIR photos are commonly used for agriculture, forestry and wetland studies because the IR band provides valuable information on vegetation health, species and biomass. Deep and clear water absorbs almost all of the NIR energy while reflecting somewhat more green and red light. Deep water free from suspended sediment will appear black. Whereas, water with substantial suspended sediment may appear in relative dark shades of blue and green.

Normal Color.

False-color Infrared using wratten #12 filter.

Multiband photography: Aerial camera using multiband photography film takes simultaneous photos in different portions of the spectrum. For example, four bands photography might include separate b/w photographs in blue, green, red and short-infrared bands.

Stereoscopic Coverage

The three-dimensional view which results when two overlapping photos (called a stereo pair), are viewed using a stereoscope. Each photograph of the stereo pair provides a slightly different view of the same area, which the brain combines and interprets as a 3-D view. Stereoscopic vision determines the distance to an object by intersecting two lines of sight.

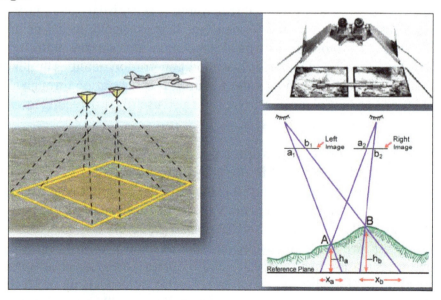

Lens Stereoscope

A lens or pocket stereoscope is a low-cost instrument that is very useful in the field as well as the office. It offers a fixed magnification, typically 2.5X. The lens stereoscope is useful for photo interpretation, control point design, and verification of mapped planimetric and topographic features.

Mirror Stereoscope

A mirror stereoscope can be used for the same functions as a lens, but is not appropriate for field use. The mirror stereoscope has a wider field of view at the nominal magnification ratio. Since photographs can be held fixed for stereo viewing under a mirror stereoscope, the instrument is useful for simple stereoscopic measurements. Mirror stereoscopes can be equipped with binocular eyepieces that yield 6X and 9X magnifications. The high magnification helps to identify, interpret, and measure photographed features.

Scale of the Photograph

The concept of scale for aerial photograph is same as that of a map. Scale is the ratio of a distance on an aerial photograph and the distance between the same two places on the ground in the real world. It can be expressed in unit equivalents like 1 cm= 1,000 km (or 12,000 inches) or as a representative fraction (1:100,000). To determine the dimension during air photo interpretation, it will be necessary to make estimates of lengths and areas, which require knowledge of the photo scale. Scale maybe expressed in three ways:

- Scale ratio: It is also referred to as the proportional scale. 1:20,000 is read as "one to twenty thousand".

- Equivalent scale: Equivalent scale is also known as the descriptive scale. For example: one inch equals 5,280 feet (1 inch = 5,280 feet).

- Graphic scale: Also called a bar scale, used on maps and drawings to represent length scale on paper with length units.

Large Scale

Larger-scale photos (e.g. 1:25000) cover small areas in greater detail. A large-scale photo simply means that ground features are at a larger, more detailed size. The area of ground coverage that is seen on the photo is less than at smaller scales.

Small Scale

Smaller-scale photos (e.g. 1:50000) cover large areas in less detail. A small-scale photo simply means that ground features are at a smaller, less detailed size. The area of ground coverage that is seen on the photo is greater than at larger scales. Following methods are used to compute scale of an aerial photograph using different sets of information:

Method 1: Scale is the ratio of the distance between two points on a photo to the actual distance between the same two points on the ground (i.e. 1 unit on the photo equals "x" units on the ground). If a 1 km stretch of highway covers 4 cm on an air photo, the scale is calculated as follows:

$$\frac{\text{Photo distance}}{\text{Ground distance}} = \frac{4\,cm}{1\,km} = \frac{4\,cm}{100000\,cm} = \frac{1}{25000}$$

So the scale is: 1/25000

Method 2: Another method used to determine the scale of a photo is to find the ratio between the camera's focal length and the plane's altitude above the ground being photographed.

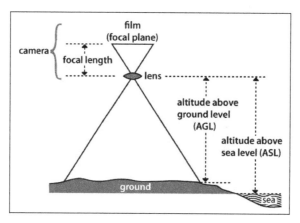

If a camera's focal length is 152 mm, and the plane's altitude Above Ground Level (AGL) is 7 600 m, using the same equation as above, the scale would be:

$$\frac{\text{Focal length}}{\text{Altitude}} = \frac{152\,mm}{7600\,m} = \frac{152\,mm}{57600000\,mm} = \frac{1}{50000}$$

So the scale is: 1/50000

Method 3: Scale of the photograph can also be calculated if we know focal length of camera and height of aircraft above the ground level.

Scale = f/H-h

Where, H = flying height of aircraft above sea level, h = height of ground above sea level and f is focal length.

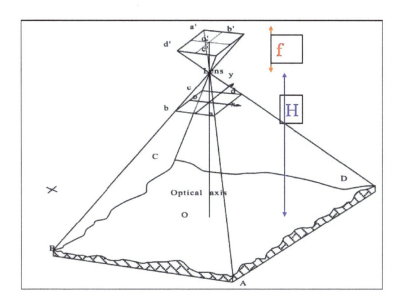

Relief Displacement

Relief displacement is the shift in an object's image position caused by its elevation above a particular datum. A vertical object (such as a building or tree) will appear to be lying along a line radial to the image nadir point. This deformation is called relief displacement.

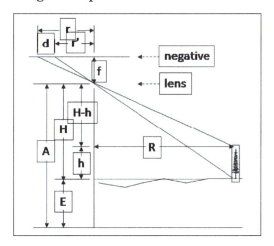

Here,

- r = Distance on the photo from the nadir to the displaced landscape feature.
- r' = Actual place on the photo where the landscape feature should be located.
- d = Relief (topographic) displacement.
- f = Focal length.
- h = Height of the landscape feature.

- A = Altitude of the aircraft above sea level.
- E = Elevation of the landscape feature.
- H = Flying height above the base of the landscape feature at nadir.
- R = Distance from the nadir to the landscape feature.

Example: Estimation of Tree Height.

- Suppose we have the measured displacement of a tree, on flat ground, or d = 2.1 mm.
- The distance from the top of the tree to the nadir of the photograph is 79.4 mm, or r = 79.4 mm.
- The flying height of the aircraft, A, above sea level is 10,000 feet.
- The elevation of the area, E, from a topographic map is 2,000 feet.
- Then what is height of the tree?

$$h = \left[\frac{(A-E)d}{r}\right]$$

$$h = \left[\frac{(10000\ feet - 2000\ feet)21.mm}{79.4}\right]$$

$$h = \left[\frac{(8000\ feet)2.1mm}{79.4\ mm}\right]$$

$$h = 211.6\ \text{feet}$$

Photo Interpretation

The identification and extraction of meaning of objects from photo is known as photo interpretation. Once corrected, and geo referenced, photos can be used for topographic mapping and as a mapping layer, with map data overlain on top. With careful interpretation, air photos are an excellent source of spatial data for studying the Earth's environment.

Photo Interpretation Equipment

Photogrammetric Workstation

Photogrammetric workstation involves integrated hardware and software systems for spatial data capture, manipulation, analysis, storage, display and output of softcopy images. These systems incorporate functionality of analytical stereo plotters, automated generation of DEM, computation of digital ortophotos, preparation of perspective views and capture @D and 3D data for use in a GIS.

Geometric Properties of Aerial Photographs

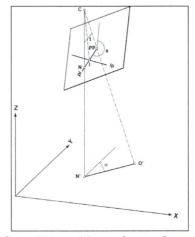

Tilted photograph in diapositive position and ground control coordinate system.

We restrict the discussion about geometric properties to frame photography, that is, photographs exposed in one instant. Furthermore, we assume central projection. Figure below shows a diapositive in near vertical position. The following definitions apply:

- Perspective center C calibrated perspective center.

- Focal length c calibrated focal length.

- Principal point PP principal point of autocollimation.

- Camera axis C-PP axis defined by the projection center C and the principal point PP. The camera axis represents the optical axis. It is perpendicular to the image plane.

- Nadir point N' also called photo nadir point, is the intersection of vertical (plumb line) from perspective center with photograph.

- Ground nadir point N intersection of vertical from perspective center with the earth's surface.

- Tilt angle t angle between vertical and camera axis.

- Swing angle s is the angle at the principal point measured from the +y-axis counter clockwise to the nadir N.

- Azimut α is the angle at the ground nadir N measured from the +Y-axis in the ground system counter clockwise to the intersection O of the camera axis with the ground surface. It is the azimut of the trace of the principal plane in the XY-plane of the ground system.

- Principal line pl intersection of plane defined by the vertical through perspective center and camera axis with the photograph. Both, the nadir N and the principal point. PP is on the principal line. The principal line is oriented in the direction of steepest inclination of of the tilted photograph.

- Isocenter I is the intersection of the bisector of angle t with the photograph. It is on the principal line.

- Isometric parallel ip is in the plane of photograph and is perpendicular to the principal line at the isocenter.

- True horizon line intersection of a horizontal plane through persepective center with photograph or its extension. The horizon line falls within the extent of the photograph only for high oblique photographs.

- Horizon point intersection of principal line with true horizon line.

Image and Object Space

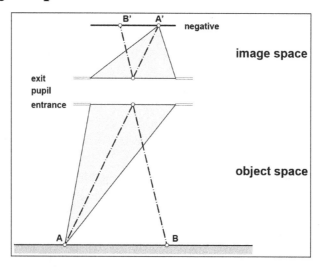

The photograph is a perspective (central) projection. During the image formation process, the physical projection center object side is the center of the entrance pupil while the center of the exit pupil is the projection center image side. The two projection centers are separated by the nodal separation. The two projection centers also separate the space into image space and object space as indicated in Figure. During the camera calibration process the projection center in image space is changed to a new position, called the calibrated projection center. This is necessary to achieve close similarity between the image and object bundle.

Photo Scale

We use the representative fraction for scale expressions, in form of a ratio, e.g. 1 : 5,000. In figure the scale of a near vertical photograph can be approximated by,

$$m_b = \frac{c}{H}$$

where m_b is the photograph scale number, c the calibrated focal length, and H the flight height above mean ground elevation. Note that the flight height H refers to the average ground elevation. If it is with respect to the datum, then it is called flight altitude H_A, with $H_A = H + h$.

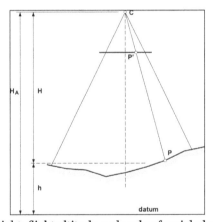

Flight height, flight altitude and scale of aerial photograph.

The photograph scale varies from point to point. For example, the scale for point P can easily be determined as the ratio of image distance CP' to object distance CP by,

$$m_b = \frac{CP'}{CP}$$

$$CP' = \sqrt{x_p^2 + y_p^2 + c^2}$$

$$CP = \sqrt{(X_P - X_C)^2 + (Y_P - Y_C)^2 + (Z_P - Z_C)^2}$$

where x_p, y_p are the photo-coordinates, X_P, Y_P, Z_P the ground coordinates of point P, and X_C, Y_C, Z_C the coordinates of the projection center C in the ground coordinate system. Clearly, above equation takes into account any tilt and topographic variations of the surface (relief).

Relief Displacement

The effect of relief does not only cause a change in the scale but can also be considered as a component of image displacement. Figure illustrates this concept. Suppose point T is on top of a building and point B at the bottom. On a map, both points have identical X, Y coordinates; however, on the photograph they are imaged at different positions, namely in T' and B'. The distance d between the two photo points is called relief displacement because it is caused by the elevation difference Δh between T and B.

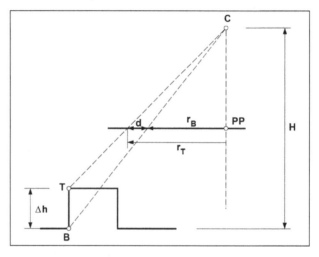

The magnitude of relief displacement for a true vertical photograph can be determined by the following equation:

$$d = \frac{r\Delta h}{H} = \frac{r'\Delta h}{H - \Delta h}$$

where $r = \sqrt{x_T^2 + y_T^2}, r' = \sqrt{x_B^2 + y_B^2}$, and Δh the elevation difference of two points on a vertical. The above equation can be used to determine the elevation Δh of a vertical object,

$$h = \frac{dH}{r}$$

The direction of relief displacement is radial with respect to the nadir point N', independent of camera tilt.

Photogeology

Interpretation of various features on aerial photograph is called as photogeology. An aerial photograph is the image of the earth surface taken from the air with the help of a camera pointing downward.

Printed Information on Aerial Photographs

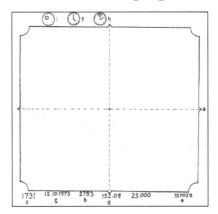

- Fiducal (collimation) mark: Fiducal mark or fiducal centre are used to identify the principal point on the photograph. Geometric centre of a photograph is called as principal point. When two lines joining opposite fiducal point intersects at a point, this point of intersection is called as principal point. There is 60% overlap of aerial photograph to the adjacent photograph. Therefore, principal point of one photograph also lies on the adjacent photographs. These points are called as 'conjugate principal point' or 'transferred principal point'. Therefore, every photograph has one principal point and one conjugate point.

- Serial number: There is a number on all photographs of same strip along the flight line is called as serial number. At the end of each flight plan serial number are recorded.

- Film (or photograph) number: Separate number is given to photographs taken from airplane is called as film number.

- Focal length (or principal distance) 'f': The distance between the lens of camera and film used is called as focal length denote by 'f'. Standard focal length of modern camera is 152mm.

- Camera number: A number is given to the camera, which is used to take the photograph. It is automatically printed on the photograph.

- Clock: To determine the speed of the aircraft time interval of two successive aircraft is used. Time when picture is taken is shown by clock.

- Date: Day, month and year are also displayed on the aerial photograph under the column of date.

- Altimeter: It gives the detail of height of aircraft from mean sea level from which photograph is captured.

- Spirit level: Tilt of photograph is shown by spirit level.

Basic Elements of Photo Interpretation

Following are the basic elements used for the identification of features on photographs and for photo interpretation:

Shape

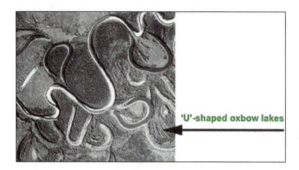

General form, structure, or outline of an object is called as shape. It is an important tool for photo interpretation. Urban or agriculture features have straight edge shape, whereas forest and other natural features have irregular shape. The shapes of objects on photographs are vertical view of the objects. It is sometimes very difficult to identify elements on ground from their vertical view. Shape of any objects act as a tool in identification of structure, composition and function of the objects. For example, an interpreter of industrial studies is not able to tell about the function of building by just seeing its front door, but with the help of vertical view of the object, he can tell more about its function on the basis of shape of the building. By seeing figure, oxbow lakes and meandering rivers are easily identified.

Size

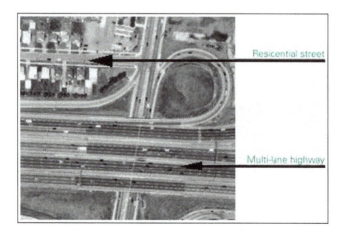

Scale of the photograph determines the size of objects on the photographs. Assessment of absolute size of the objects and their relative size to other object aids in photo interpretation. For instance, it helps in determining if the object is small pond or a large lake. It also helps in differentiating smaller roads from larger highways, and also helps in distinguishing between smaller tributaries from large river. By quick estimation of size of the object can drive to interpretation to a suitable result easily. For instance, if an interpreter had to identify land use zones, and if there is an area with a number of buildings in it, large buildings like factories or warehouses can be commercial property, while smaller buildings can be of residential use. Residential Street and multi-lane highway can be easily demarcated on figure.

Shadows

Size and shape of any object can be assumed by their shadow. Therefore, it is also useful in interpretation. It can give a clue related to the profile and relative height of an object or targets that helps in easier identification of objects. It is also very useful most particularly in radar imagery for identification of topography and landforms on the earth surface. However, some useful information is also lost because of shadows as objects are very less detectable in the area of influence of shadow. High rise building in figure can be identifies by seeing its large shadow in top of the image in its right hand side and low rise buildings are seen in bottom left part of the image.

Tone

Relative whiteness and blackness of the photograph is referred as tone and it is the outcome of the reflectance of light by an object. In black and white photograph, tone is used as fundamental tool for photo interpretation. There are many factors, which influence the tones of photographic images. These factors are angle of sun, number of wave reflected by the object etc. Smooth surface reflect more light so they appear light in tone on photograph, for instance a black asphalt road appear light in tone. Tonal variation is used in variation field like soil scientist used it for soil classification; the foresters used it to differentiate hardwood from coniferous forest and geologist for mapping of minerals, lithology and classification of rocks. Light tone is sand and dark tone is water can be easily differentiated in figure.

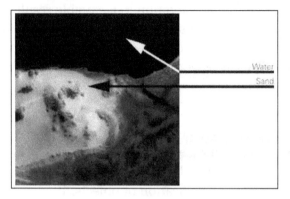

Colour

When any object reflects light in a particular wavelength they are showing different shades of colour. For instance, naturally, vegetation appears green because they reflect a larger part of green light than the other colour like blue or red. Colour is an important tool for photo interpretation because, it is easier for humans to differentiate different colour than different shades of grey. So in coloured photograph greater amount of information can be extracted than the panchromatic photograph. False colour composition (FCC) developed during World War II is very useful for the study of conditions of plants, distribution of vegetation, drainage delineation and assessment of soil-moisture.

Texture

Texture is defined in terms of "smoothness" and "roughness". The amount of change in photograph causes smoothness or roughness. Object, which has less variation, appears smooth and object, which has more variation, appears rough. Generally smother surface has lighter tone in the photograph. Rough texture like tall grasses and shrubs are of dark tone. For example, features like grass, cemented features and water appear smooth, while canopy covers of forest appear rough. Texture of forest appears rough and texture of calm water id smooth as shown in figure.

Patterns

Pattern is an important clue for identification of features. Smallest and significant patterns can be captured with the help of aerial photograph. There can be some natural and some cultural pattern and some are resulted because of interaction of man and nature. Pattern is nothing but spatial arrangement of phenomena on the earth surface. This spatial arrangement is used in the identification of objects. For example, unmanaged area of trees have random pattern whereas orchard are evenly spaced arrangement of trees. Natural forest, plantation and open field can be identified in figure on the basis of their pattern.

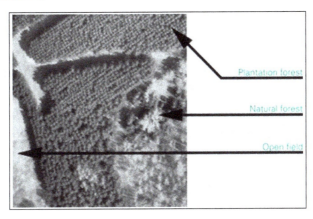

Relationship or Association

Some type of association always exists between some groups of objects. The background of an object can endow with handy information about the object. For example, in the setting of single family housing it is not permissible to have a nuclear power plant. Green area in urban environment must be a park or a cemetery. Wetlands may be located next to rivers, lakes, or estuaries. Commercial centers will likely be situated along major roads, railroads, or waterways. In figure, one can easily identify railway track along the coast, dry dock and ship along the coast and the railway track and water harbour along the railway tracks and along the ship.

Applications of Aerial Photography

In Archaeology

In archaeology aerial photography is ideal for locating lost monuments and tracking features, especially those that are not visible at ground level, those that are under the soil and cannot be seen on a field walk and those that can only be seen under certain conditions. They are usually discovered through any of the following:

- Crop Marks and Parch Marks: Seen in summer, crop marks are signs of a subterranean feature that show up as irregularities in the pattern of crops. Growth of the crop might be stunted due to extant remains such as stone foundations, or they might be higher than the surrounding crop due to underlying water systems such as dried up drainage channels or long-gone artificial water features such as fishponds. Parch marks occur in areas of particularly dry summer. In some conditions, the crop may simply be a different colour. Parch marks differ in that they are discolorations in the crop as a result of prolonged drought. Areas where ground water dries up quickly and areas where there may be more groundwater will show up clearly. Caution is advised when interpreting both crop marks and parch marks as the anomalies may be archaeological, geological, or due to variations in soil and ground water courses. Modern pipes may also flag a false positive for features of interest.

- Soil Marks: Best studied in winter when no crops are growing or grasses have large died off, both rainy and dry conditions are conducive to picking out buried features. Typically showing up as darker areas, they can indicate underlying stonework, the outline of prehistoric features such as barrows and cursus monuments, and ditches. The same issues above apply - they could be natural or modern features.

- Low Profile Monuments: From the ground they may seem like natural bumps in the ground or be so slight as to be barely perceptible. From the air, their appearance is far more revealing. On their own they may or may not look like anything important but if accompanied with the above, can appear more significant.

In Urban Studies

Urban development and the history of urbanism is a growing niche of landscape studies which has a wide range of uses through history and archaeology, the history of cartography, the history of commerce, sociology and even for modern urban planning. Town developers need to study the impact of expansion and development of urban centres on the landscape and the impact on the environment. New facilities (for example a new sports stadium) will require a rethink of the infrastructure and the impact that the new facility will have on people living in the area - will we need to build more houses? Upgrade the roads? Will this affect protected areas? Aerial photography taken at low levels is vital to examining the existing infrastructure.

In Climate Change

We all know about the effects of climate change on global temperatures. These global changes are reflected everywhere, and societies and communities are seeing changes to their local environment. If it isn't river beds drying up, droughts getting longer, wetter seasons getting wetter and the reduction of inland lakes drying up completely, one of the most practical applications is tracking of invasive species into water bodies that just a few years ago would not have provided an adequate environment for those species. Researchers keep vital records in changes over seasons and years to track local effects of climate change and risks to local ecosystems. Localised aerial photographs will highlight the die-off of certain vegetation, or the increase of invasive species.

In Other Earth Sciences

They can also be used to study the process of natural changes, such as variations in soil and geology over time as well as changes to the underlying ground that leads to disasters such as landslides. Not quite as useful to geologists due to the relative expense and difficulty in interpretation compared to archaeological applications, aerial survey nevertheless has uses and benefits and the historical record for changes to the natural landscape is vital to understanding how the landscape may change in future. Annual rainfall, whether lower or higher than normal, can have far-reaching consequences and it is this where geology's interests in aerial photography are most important.

Though increasingly taken over by satellite images and digital mapping of GIS in recent years, photogeology still has some practical applications for finding mineral and fuel deposits, mapping areas and tracking geological changes and water management as well as general geological research that other applications cannot contribute to. A great example of this is water drainage ahead of proposed new urban developments - flood plain risks and subsidence.

Platforms and Sensors

Platform is a stage where sensor or camera is mounted to acquire information about a target under investigation. A platform is a vehicle, from which a sensor can be operated. Platforms can vary from stepladders to satellites. There are different types of platforms and based on its altitude above earth surface, these may be classified as:

Ground based Platforms

Wide varieties of ground-based platforms are used in remote sensing. Some of the common ones are hand held devices, tripods, towers and cranes. To study properties of a single plant or a small patch of grass, ground based platform is used. Ground based platforms (hand-held or mounted on a tripod) are also used for sensor calibration, quality control and for the development of new sensors.

For the field investigations, some of the most popular platforms have been used are 'cherry picker platform, portable masts and towers. The cherry picker platforms can be extended to approx. 15m. They have been used by various laboratories to carry spectral reflectance meters and photographic systems. Portable masts are also available in various forms and can be used to support cameras and sensors for testing. The main problem with these masts is that of stabilizing the platform, particularly in windy conditions. Permanent ground platforms like towers and cranes are used for monitoring atmospheric phenomenon and long-term monitoring of terrestrial features. Towers can be built on site and can be tall enough to project through a forest canopy so that a range of measurements can be taken from the forest floor, through the canopy and from above the canopy.

Balloon Platforms

Balloons as platforms are not very expensive like aircrafts. They have a great variety of shapes, sizes and performance capabilities. The balloons have low acceleration, require no power and exhibit low vibrations. There are three main types of balloon systems, viz. free balloons, Tethered balloons and Powered Balloons. Free balloons can reach almost top of the atmosphere; hence, they can provide a platform at intermediate altitude between those of aircraft and spacecraft.

Free floating or anchored balloons have altitude range of 22-40 km and can be used to a limited extent as a platform. It is used for probing the atmosphere and also useful to test the instrument under development.

Aircraft Platform

Aerial platforms are primarily stable wing aircraft. Helicopters are also occasionally used for this purpose. Generally, aircraft are used to collect very detailed images. Helicopters can be for pinpoint locations but it vibrates and lacks stability.

- Low Altitude Aircraft: It is most widely used and generally operates below 30,000 ft. They have single engine or light twin engine. It is suitable for obtaining image data for small areas having large scale.

Low altitude aircraft produced image.

- Low Altitude Aircraft: It is more stable and operates above 30,000 ft. High altitude aircraft includes jet aircraft with good rate of climb, maximum speed, and high operating ceiling. It acquires imagery for large areas (smaller scale). Examples are NHAP, NAPP, AVIRIS. Aircraft platform acquire imagery under suitable weather conditions. It controls platform variables such as altitude. Time of coverage can also be controlled. However, it is expensive, less stable than spacecraft and has motion blurring.

Rockets as Platforms

Prior to use of airplanes, aerial photographs were obtained by rocketing a camera into the sky and then retrieving the camera and film. Synoptic imagery can be obtained from rockets for areas of some 500,000 square km. The Skylark earth Resource Rocket is fired from a mobile launcher to altitudes between 90 - 400 kms. With the help of a parachute, the payload and the spent motor are returned to the ground gently thereby enabling speedy recovery of the photographic records. This rocket system has been used in surveys over Australia and Argentina. In 1946, V-2 rockets acquired from Germany after World War II were launched to high altitudes from White Sands, New Mexico. These rockets contained automated still or movie cameras that took picture as the vehicle ascended. The main problem with rockets is that they are one-time observations only. Except for one time qualitative or reconnaissance purposes, rocket platforms are not of much use in regular operational systems.

Spacecraft as Platform

Remote sensing is also conducted from the space shuttle or artificial satellites. Artificial satellites are manmade objects, which revolve around another object. The 1960s saw the primary platform used to carry remotely sensed instruments shifted from airplanes to satellite. Satellite can cover much more land space than planes and can monitor areas on a regular basis.

Beginning with the first television and infrared observation Satellite (tiRoS1) in 1960, early weather satellites returned rather poor views of cloud patterns and almost indistinct images of the earth's surface. Space photography becomes better and was further extended with the Apollo program. Then in 1973, Skylab the first American space workshop was launched and its astronauts took over 35,000 images of the earth with the earth Resources experiment Package (eReP) on board. Later on with LANADSAT and SPOT Satellite program, space photography received a higher impetus.

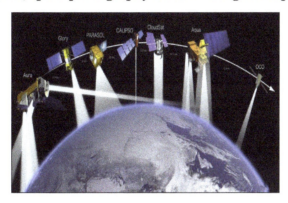

Types of Orbits

Sun-synchronous Orbits

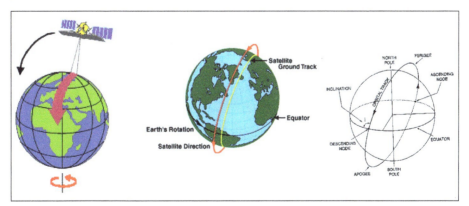

In sun-synchronous orbit, satellite is orbiting in synchronous with sun. In this type of orbit, satellite passes every location at the same time each day. Such an orbit is near polar with altitudes between 300 and 1000 km. They are also known as "polar orbiters"

because of the high latitudes they cross. The noon satellites pass over near noon and midnight, whereas morning satellites pass over near dawn and dusk.

Satellite in a sun-synchronous orbit has an inclination that carries the satellite track westward at a rate that compensates for the change in local sun time as the satellite moves from north to south. In fact most earth observation satellites are placed in orbits designed to acquire imagery between 9.30 and 10.30 A.M. This local time is set to minimize cloud cover in tropical regions and provides optimum illumination. Satellites in sun-synchronous orbits pass from north to south on the sunlit side (descending node) and from south to north on the shadowed side (the ascending node). During the descending pass, sensors that depend on reflected solar radiation acquire data, but radar and thermal sensors can acquire data independently of solar illumination, thereby observing the earth's surface during both passes.

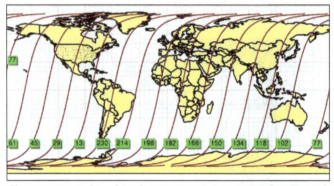

Sun-synchronous satellites: 700-900 km altitude, rotates at circa 81-82 degree angle to operator: capture imagery approx. the same time each day (10am =/-30 minutes).

The major advantage of such an orbit is that data taking is standardized with respect to the sun angle. Secondly, since the sun synchronous orbit is near polar, it provides nearly global coverage. Through these satellites, the entire globe is covered on regular basis and gives repetitive coverage on periodic basis. Some of them sun-synchronous satellites are LANDSAT Series, SPOT Series, IRS Series, NOAA, SEASAT, TIROS.

Geostationary Orbits

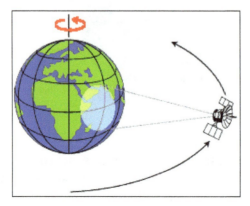

In geosynchronous orbit, satellite in synchronous with earth. It is basically designed to maintain a constant position with respect to a specific portion of the earth's surface area day and night. These satellites move along in the same direction of the rotation of the earth, taking the same amount of time as the earth's rotation. Its coverage is limited to 70 °N to 70 °S latitudes and one satellite can view one-third globe. While moving west to east such satellites must have a velocity equal to that of the earth rotating about its axis west to east.

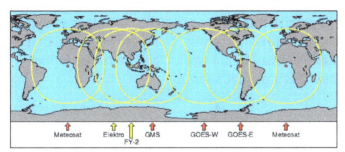

Global Geostationary Satellite Coverage.

The main advantage of this satellite is that since it is stationary (rather appears stationary); it allows continuous viewing of that portion of the earth within the line of sight of the satellite sensors. However, satellite can only "see" one hemisphere. These are orbiting approximately 36,000 km above the Earth. Geostationary orbits are ideal for meteorological or communications satellites. Some examples of geostationary satellites are GOES, METOSAT, INTELSAT, INSAT.

Sensors

Sensor is a device that gathers energy (EMR or other), converts it into a signal and presents it in a form suitable for obtaining information about the target under investigation. Remote sensors are mechanical devices, which collect information, usually in storable form, about objects or scenes, while being at some distance from them. Sensors used for remote sensing can be either those operating in Optical Infrared (OIR) region or those operating in the microwave region. Depending on the source of energy, sensors are categorized as active or passive sensors.

Active Sensors

Active sensors are those, which have their own source of EMR for illuminating the objects. Radar (Radio Detection and Ranging) and Lidar (Light Detection and Ranging) are some examples of active sensor. Photographic camera becomes an active sensor when used with a flash bulb. Radar is composed of a transmitter and a receiver. The transmitter emits a wave, which hits objects in the environment and gets reflected or echoed back to the receiver. The main advantage is that active sensors can obtain imagery in wavebands where natural signal levels are extremely low and also are independent of natural illumination. The major disadvantage with active sensor is that it needs high energy levels, therefore adequate inputs of power is necessary.

Passive Sensors

Passive sensors do not have their own source of energy. These sensors receive solar electromagnetic energy reflected from the surface or energy emitted by the surface itself. Therefore, except for thermal sensors they cannot be used at night time. Thus in passive sensing, there is no control over the source of electromagnetic radiation. Photographic cameras (without the use of bulb), multispectral scanners vidicon cameras etc. are examples of passive remote sensors. The advantage with passive sensor is that it is simple and do not require high power. The disadvantage is that during bad weather conditions the passive sensors do not work. The Thematic Mapper (TM) sensor system on the Landsat satellite is a passive sensor.

Types of Scanner

Whiskbroom Scanner

A transverse scanning system is an electro-mechanical device (small number of sensitive diodes) that obtains data from narrow swaths of terrain right angle to the direction of flight. It means scanner sweeps perpendicular to the path or swath, centered directly under the platform, i.e. at 'nadir'. The forward movement of the aircraft or satellite allows the next line of data to be obtained in the order 1, 2, 3, 4 etc. In this way, an image is built up in a sequential manner. In whiskbroom system, scanning is done by using rotating or oscillating mirror e.g. LANDSAT MSS/TM.

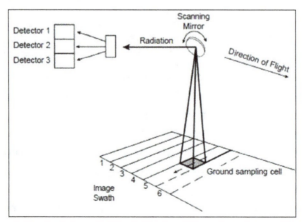

Pushbroom Scanner

The pushbroom scanner has a linear array of detectors, in which each detector measures the radiation reflected from a small area on the ground. In this type of scanning system, linear array of detectors scan in the direction parallel to the flight line. Linear arrays normally consist of numerous charge coupled devices (CCDs) positioned end to end. Charge-Coupled Devices are designed to be very small and a single array may contain over 10,000 individual detectors. Normally, the arrays are located in the focal plane of the scanner such that all scan lines are viewed by all arrays simultaneously. This system is more reliable as against the scanning mirror of the transverse system and the detectors are light and need little power to operate.

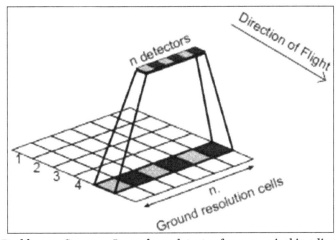
Pushbroom Scanner: It employ a detector for every pixel in a line.

Different Sensors and their Characteristics

Optical Infrared Sensors

Optical infrared remote sensors are used to record reflected/emitted radiation of visible, near middle and far infrared regions of electromagnetic radiation. They can observe for wavelength extended from 400-2000 nm. Sun is the source of optical remote sensing. There are two kinds of observation methods using optical sensors: visible/near infrared remote sensing and thermal infrared remote sensing.

Visible/Near Infrared Remote Sensing

In this, visible light and near infrared rays of sunlight reflected by objects on the ground is observed. The magnitude of reflection infers the conditions of land surface, e.g., plant species and their distribution, rivers, lakes, urban areas etc. In the absence of sunlight or darkness, this method cannot be used.

- Panchromatic Imaging System: In this type of sensor, radiation is detected within a broad wavelength range. In panchromatic band, visible and near infrared are included. The imagery appears as a black and white photograph. Examples of panchromatic imaging system are Landsat ETM+ PAN, SPOT HRV-PAN and IKONOS PAN, IRS-1C, IRS-1D and CARTOSAT-series. Spectral range of Panchromatic band of ETM+ is 0.52 µm to 0.9 µm, CARTOSAT-2B is 0.45-0.85 µm, SPOT is 0.45-0.745 µm.

- Multispectral imaging system: The multispectral imaging system uses a multi-channel detectors and records radiation within a narrow range of wavelength. Both brightness and color information are available on the image. LANDSAT, LANDSAT TM, SPOT HRV-XS and LISS etc. are the examples.

Thermal Infrared Remote Sensing

In thermal infrared remote sensing, sensors acquire those energy/heat that are radiated by earth surface due to interaction with solar radiation. This is also used to observe the high temperature areas, such as volcanic activities and forest fires. Based on the strength of radiation, one can surface temperatures of land and sea, and status of volcanic activities and forest fires. This method can observe at night when there is no cloud. The optical remote sensing can be classified into following:

Hyperspectral Imaging System

Hyperspectral imaging system records the radiation of terrain in 100s of narrow spectral bands. Therefore the spectral signature of an object can be achieved accurately, helps in object identification more precisely. Example, Hyperion data is recorded in 242 spectral bands and AVIRIS data is recorded in 224 spectral bands.

Microwave Sensors

These types of sensors receive microwaves, which are having longer wavelength than visible light and infrared rays. The observation is not affected by day, night or weather. Microwave portion of the spectrum includes wavelengths within the approximate range of 1mm to 1m. The longest microwaves are about 2,500,000 times longer than the shortest light waves. There are two types of observation methods using microwave sensor:

- Active sensor: The sensor emits microwaves and observes microwaves reflected by land surface features. It is used to observe mountains, valleys, surface of oceans wind, wave and ice conditions.

- Passive sensor: This type of sensor records microwaves that naturally radiated from earth surface features. It is suitable to observe sea surface temperature, snow accumulation, thickness of ice, soil moisture and hydrological applications etc. RISAT is an Indian remote sensing satellite provides microwave data.

Characteristics of Some Scanners

LANDSAT

NASA, with the co-operation of the U.S. Department of Interior, began a conceptual study of the feasibility of a series of Earth Resources Technology Satellites (ERTS). During 1975, NASA officially renamed the ERTS programme as the "LANDSAT" programme. There have been four different types of sensors included in various combinations on these missions. These are Return Beam Vidicon camera (RBV) systems, Multispectral Scanner (MSS) systems, Thematic Mapper (TM) and Enhanced Thematic Mapper (ETM).

Table: Landsat 3 RBV Specifications.

Satellite	Sensor	Band No.	Wavelength (µm)	Geo. Resolution
Landsat 3	RBV	1	0.505 to 0.75	40 x 40m

Table: Landsat 4-5 MSS and TM specification.

Sensor		Band No.	Wavelength (µm)	Geo. Resolution
MSS	Green Visible	1	0.50 to 0.60	82m
o	Red Visible	2	0.60 to 0.70	82m
o	Near IR	3	0.70 to 0.80	82m
o	Near IR	4	0.80 to 1.10	82m
o	Blue Visible	1	0.45 to 0.52	30m
o	Green Visible	2	0.52 to 0.60	30m
o	Red Visible	3	0.63 to 0.69	30m
o	Near IR	4	0.76 to 0.90	30m
o	Mid IR	5	1.55 to 1.75	30m
o	Thermal IR	6	10.4 to 12.5	120m
o	Mid IR	7	2.08 to 2.35	30m

Table: Landsat 7 ETM+ characteristics.

Sensor		Band No.	Wavelength (µm)	Geo. Resolution
ETM+	Blue Visible	1	0.45 to 0.52	30m
o	Green Visible	2	0.52 to 0.60	30m
o	Red Visible	3	0.63 to 0.69	30m
o	Near IR	4	0.76 to 0.90	30m
o	Mid IR	5	1.55 to 1.75	30m
o	Thermal IR	6	10.4 to 12.5	60m
o	Mid IR	7	2.08 to 2.35	30m
o	Panchromatic	8	0.50 to 0.90	15m

SPOT

The SPOT (Satellites Pour l'Observation de la Terre or Earth-observing Satellites) remote-sensing programme was set up in 1978 by France in partnership with Belgium and Sweden. The SPOT satellites constellation offers acquisition and revisit capacity allowing to acquire imagery from anywhere in the world, every day. Each SPOT payload comprises two identical high-resolution optical imaging instruments, which can operate simultaneously or individually in either panchromatic (P mode: a single wide band in the visible part of the spectrum) or multispectral mode (XS mode: the green, red, and infrared bands of the electromagnetic spectrum). The temporal resolution is shortened from 26 to 4-5 days for the temperate zones.

Satellites SPOT 4 and 5 together ensure the provision of simultaneous acquisition of stereo pairs, high-resolution SPOT images and of VEGETATION global images. The continuity of the SPOT programme is planned with the development of Spot 6 and 7, which will offer 2-meter resolution images in a 60 km by 60 km swath.

SPOT 1, 2 and 3 Satellite's Characteristics

The SPOT 1, 2 and 3 satellites were identical and their payloads consisted of two identical HRV (Visible High-Resolution) optical instruments, data recorders (on magnetic tapes), and a system for transmitting the images to the ground-based receiving stations.

Mode	Band	Spectral band	Resolution
Mode	Band	Spectral Band	Resolution
XS-multispectral	XS1	0,50 - 0,59 µm (green)	20m × 20m
	XS2	0,61 - 0,68 µm (red)	20m × 20m
	XS3	0,78 - 0,89 µm (near IR)	20m × 20m
P-panchromatique	PAN	0,50 - 0,73 µm	10m × 10m

SPOT 5

The main payload consists of high resolution imaging instruments delivering the following product improvements compared to Spot 4: The HRS (High-Resolution Stereoscopic) imaging instrument dedicated to taking simultaneous stereopairs of a swath 120 km across and 600 km long; A ground resolution of 5 and 2.5 meters in panchromatic mode; A resolution in multispectral mode of 10 m in all 3 spectral bands in the visible and near infrared ranges. The spectral band in the short wave infrared band (essential for VEGETATION data) is maintained at a resolution of 20 m due to limitations imposed by the geometry of the CCD sensors used in this band.

The Spot 5 spectral bands are the same as those for Spot 4. The panchromatic band does, however, return to the values used for Spot 1-2-3. As requested by many users, this ensures continuity of the spectral bands established since Spot 1.

- Altitude: 822 km.
- Inclination: 98.7 degrees.
- Orbit: Sun-synchronous polar.
- Period of revolution: 101 minutes.
- Swath width: 60 × 60 to 80 km.
- Repeat cycle: 26 days.
- Satellite: SPOT 5 (04/05/2002 – still operational).

High Resolution Geometric Sensors

Two HRG instruments are capable of generating data at 4 resolution levels with the same 60 km swath.

Mode	Band	Spectral band	Resolution
Multispectral	B1	0,50 - 0,59 µm	10m x 10m
	B2	0,61 - 0,68 µm	10m x 10m
	B3	0,78 - 0,89 µm	10m x 10m
	SWIR	1,58 - 1,75 µm	20m x 20m
M - monospectral	PAN	0,51 - 0,73 µm	5m x 5m (or 2.5m x 2.5m in super mode)

High-Resolution Stereoscopic Sensors

The ability to acquire stereopair images quasi-simultaneously (90 sec) is a considerable advantage for the quality of digital elevation model (DEM) production. The resemblance between the two images is indeed maximum. Characteristics of sensor are given in the table below:

Mode	Band	Spectral band	Resolution	Swath	Max scene length	Viewing angle of the telescopes
M - monospectral	PAN	0,51 - 0,73 µm	10m × 10m	120 km	600km	± 20°

Vegetation Sensor

The vegetation programme is co-financed by the European Union, Belgium, France, Italy, and Sweden and being conducted under the supervision of the CNES (National Centre for Space Studies, France). The aim of the vegetation instrument is to provide accurate measurements of the main characteristics of the Earth's plant cover. Practically daily global coverage and a resolution of 1 km make this sensor an ideal tool for observing long-term regional and global environmental changes.

Vegetation works independently from the HRVIRs. It includes a wide-angle radiometric 'camera' operating in four spectral bands (blue, red, near infrared, and middle-infrared). Given its 2,250km swaths, this instrument is thus able to cover almost all of the Earth's dry land in just one day.

Band	Spectral band	Resolution	Applications
B0	0,43 - 0,47µm (blue)	1165m × 1165m	Oceanographic applications/ Atmospheric corrections
B2	0,61 - 0,68 µm (red)	1165m × 1165m	Vegetation photosynthesis activity
B3	0,79 - 0,89 µm (near IR)	1165m × 1165m	
MIR	1,58 - 1,75 µm (middle IR)	1165m × 1165m	Ground and vegetation humidity

Image Rectifications

Remote sensing data acquired from imaging sensors mounted on a satellite or any other aerial platform may contain various kinds of distortions and deficiencies. These may include geometric distortions or radiometric distortions. Correction or removal of both of these forms of distortions is required before using the remote sensing data for image interpretation. Pre-processing of remotely sensed data is therefore carried out to rectify and to restore the image by removing such distortions and deficiencies. The pre-processing may be broadly classified as geometric corrections and radiometric corrections.

Geometric Corrections

Geometric distortions may render the remote sensing data unusable for several purposes unless some preprocessing is carried out. These distortions may occur due to various reasons including motion of the sensor platform, relief displacement, and curvature of the earth, rotation of the earth, atmospheric refraction and nonlinearities in the sweep of a sensor's IFOV (instantaneous field of view). Some of these distortions may be systematic which may be predictable. On the other hand, some of these errors are random or non-systematic in nature.

Systematic distortions are also known as internal distortions, which may be due to the geometry of the sensors. Systematic errors or distortions may be easily determined and removed by applying suitable corrections based on relationships that are mathematically derived through modeling the sources of errors. Whereas, nonsystematic distortions (external distortions) may be due to attitude of the sensor (including roll, pitch and yaw) or due to changes in altitude of the scanners. Correction to these errors can be applied using the knowledge of platform ephemeris; ground control points (GCPs) etc.

Preprocessing of the raw data obtained from satellites is usually required for eliminating different kinds of distortions in the data. This is carried out to precisely match the position of individual pixels in the remote sensing image with their corresponding locations on the surface of earth. The purposes of geometric corrections applied on remote sensing include:

- Transformation of an image corresponding to a map projection.
- Identification of points of interest on the image and on map.
- Preparation of mosaics of images corresponding to their geographical positions.
- Overlay analysis of various images acquired at different time or by different sensors.
- Incorporation of remote sensing data into Geographic Information System (GIS).

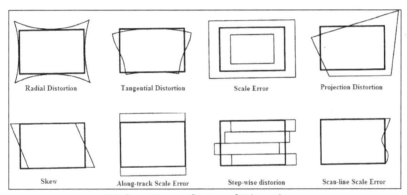
Various Types of Internal Distortions.

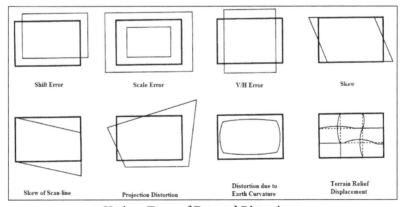
Various Types of External Distortions.

Systematic Distortions

Various types of systematic distortions are introduced in the remote sensing data due to various factors including scan skew, variation in platform velocity, earth rotation, variation in mirror scan velocity, panoramic, perspective and aspect ratio.

- Scan Skew: Scan skew in the image is produced if the ground swath is slightly skewed with respect to the ground track i.e. it is not normal to the ground track. It generally happens due to the forward motion of the platform during the time taken by each mirror sweep. As the satellite makes a move along a fixed orbital path, the Earth below it moves/rotates on its axis from west to east. Geometry of the satellite image is thus skewed due to the relative position of the fixed path of the satellite with that of the rotation of the earth about its axis. Thus, the area covered by each optical sweep of the image sensor is slightly towards the west of its previous sweep, resulting in the skew in the image. The correction to skew distortion involves offsetting each successive scanline slightly towards the west.

- Platform Velocity: The variation in the velocity of the platform results in the variation of the ground track coverage in successive mirror scans. This produces along-track scale distortion in the scanned data.

- Earth Rotation: The scanning of the ground track by a sensor takes a significant time period during which the Earth also rotates by a significant amount. This causes a shift of the ground swath being scanned and thus resulting in a long-scan kind of distortion.

- Mirror-Scan Velocity Variance: The rate of mirror scanning also varies usually across a given scan due to variation in the mirror scan velocity. This also results in along-scan type of geometric distortions in the remote sensing data.

- Panoramic Distortion: The ground area imaged by a sensor is not proportional to the scan angle rather it is proportional to the tangent of the scan angle. It also introduces along-scan distortions.

- Perspective: Scanned data is also subjected to long-scan distortions if the images are used to represent the projection of a point on the earth, on a plane that is tangent to the surface of the earth, with all the projection lines being normal.

- Aspect Ratio: Some of the sensors, e.g. Landsat MSS, produce images containing non-square pixels. It may result in due the difference between the instantaneous field of view and the spacing between pixels along each scan line. The oversampling in the cross-track direction, leads to the creation of pixels that are not square. Knowing the correct spacing between scan lines and that between pixels along a scan line, the aspect ratio of pixels may be corrected.

For the given ephemeris of a satellite and geometry of the sensors, geometric distortions may be computed and accordingly correction to the scanned data may be applied. Thus, geometric or systematic distortions may be corrected using a derived mathematical model.

Non-systematic Distortions

These are external geometric errors, which are usually caused by various spatial and temporal variations in different phenomena. These may include random movements of the platform at the time of scanning. Non-systematic distortions are usually introduced due to: Altitude changes and Attitude changes (roll, pitch, and yaw).

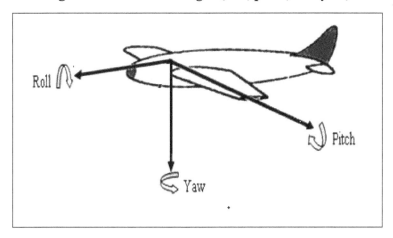

Correction to these kinds of distortions includes use of various mathematical models as well as the use of ground control points (GCP). Position of about Ground control points may be collected from topographic maps or other sources e.g. global positioning system (GPS). The method of correcting the digital images using GCPs is termed as "Image-to-ground geo correction" or "Image-to-map rectification". Another technique known as "image-to-image correction" is also commonly used. It involves fitting of the coordinate system of one digital image to that of a second image of the same area.

Coordinate Transformation

The technique of coordinate transformation is useful in carrying out geometric corrections with the help of ground control points. In this process, rearrangement of pixels in the raw image is carried out over a new grid system. Selection of suitable transformation formula as well as suitable ground control points is important in carrying out coordinate transformation.

Transformation Equation

The formula used for coordinate transformation depends upon nature of geometric distortions. Use of polynomial equation is common in computing the rectified pixel coordinates. A first order transformation may correct scale, skew, and rotation in an image. Second order transformation or higher nonlinear transformations may be used to correct nonlinear distortions including the distortions caused by curvature of the earth as well as camera lens distortion. In general, a third order polynomial is considered to provide sufficient corrections for various existing remote sensing systems including

LANDSAT. Various forms of polynomials used in geometric corrections include the following:

- Linear Polynomial (First Order):

$$x = a_0 + a_1 X + a_2 Y$$
$$y = b_0 + b_1 X + b_2 Y$$

- Quadratic Polynomial (Second Order):

$$x = a_0 + a_1 X + a_2 Y + a_3 XY + a_4 X^2 + a_5 Y^2$$
$$y = b_0 + b_1 X + b_2 Y + b_3 XY + b_4 X^2 + b_5 Y^2$$

- Cubic Polynomial (Third Order):

$$x = a_0 + a_1 X + a_2 Y + a_3 XY + a_4 X^2 + a_5 Y^2 + a_6 X^2 Y + a_7 XY^2 + a_8 X^3 + a_9 Y^3$$
$$y = b_0 + b_1 X + b_2 Y + b_3 XY + b_4 X^2 + b_5 Y^2 + b_6 X^2 Y + b_7 XY^2 + b_8 X^3 + b_9 Y^3$$

A three dimensional (3-D) generalized form of polynomials may be expressed as:

$$x = \sum_{i=0}^{m} \sum_{j=0}^{n} \sum_{k=0}^{p} a_{ijk} X^i Y^j Z^k \qquad y = \sum_{i=0}^{m} \sum_{j=0}^{n} \sum_{k=0}^{p} b_{ijk} X^i Y^j Z^k$$

where, a and b are model parameters; X, Y, Z are the cartographic/terrain coordinates; and m, n, and p are integer values (generally between 0 and 3). The sum of m, n, and p – generally being equal to 3 – represents the order of polynomial equation.

Selection of GCPs

Number of ground control points and selection of their location play crucial role in the accuracy of geometric correction. Number of GCPs required for this purpose depends upon the type of formula/polynomial used for geometric correction. Total number of GCPs should be more than the number of unknown parameters in the polynomial equation used. The minimum number of GCPs required for geometric correction using a polynomial is given by the following formula:

$$n_{min} = \frac{(t+1)(t+2)}{2}$$

Where, n_{min} is the minimum number of GCPs required, and t is the order of polynomial used in geometric correction. However, more number of GCPs may be selected depending upon the topography of the area, accuracy required in geometric correction and type of formula used. Selection of GCP for geometric corrections should be such that they are almost equally spaced over the entire image including the corner areas.

Resampling

Resampling is carried out after the process of transformation is completed. It is a process that involves matching coordinates of image pixels to their corresponding position on the surface of earth and creates a new image on a pixel-by-pixel basis. Often, a mismatch occurs between the grid of pixels in the source image and that in the reference image. Therefore, pixels are resampled so that new data file values can be computed for the output file. Various methods commonly used for resampling include nearest neighborhood, bilinear and cubic interpolations.

Nearest neighborhood, Resampling is one of the fastest methods of resampling. It is also sometimes termed as zero-order interpolation. In this approach, the value of each new pixel in the resampled image is assigned based upon the value of pixel nearest to it. The advantage of this system is that no calculation is required to determine the output pixel value, once its location has been calculated. Whereas, other methods tend to determine the pixel value based upon averaging of values of surrounding pixels. Nearest neighborhood, method is easy to use, for this purpose, as compared to other interpolation methods. However, the output images obtained through this method have a rough appearance as compared to the original image. Further, there may be instances of loss of data values and duplication of values of some pixels. The loss in data may lead to break in appearance of various linear features such as roads, railways, canals, streams.

Bilinear Resampling is also known as first-order resampling method in which, values of four pixels surrounding the pixel of interest are used to determine its values. Bilinear resampling produces smoother images as compared to that obtained through nearest neighborhood method. However, this is a slow process as extra computation efforts are to be applied in this method. Moreover, the original data of the image is altered in the output image and contrast is also reduced.

In Cubic (Second-order) Resampling method, a 4×4-pixel neighborhood is used for computation of pixel values in the resampled image. With the use of this method, the smoothening effect as produced by bilinear resampling may be avoided. However, even more computational efforts are required in this method. Further, this method renders overshoot of pixel values on either side of sharp edges.

Image Enhancement

Image enhancement is the improvement of satellite image quality without knowledge about the source of degradation. Many different, often elementary and heuristic methods are used to improve images in some sense.

Image restoration removes or minimizes some known degradations in an image. In many image processing applications, geometrical transformations facilitate processing.

Examples are image restoration, where one frequently wants to model the degradation process as space-invariant, or the calibration of a measurement device, or a correction in order to remove a relative movement between object and sensor. In all cases the first operation is to eliminate a known geometrical distortion.

The geometric rectification imagery has to be enhanced to improve the effective visibility. Image enhancement techniques are usually applied to remote sensing data to improve the appearance of an image for human visual analysis. The main focus of enhancement methods follows these procedures in to image segmentation, clustering and geometric transformations.

Apart from geometrical transformations some preliminary grey level adjustments may be indicated, to take into account imperfections in the acquisition system. This can be done pixel by pixel, calibrating with the output of an image with constant brightness. Frequently space-invariant grey value transformations are also done for contrast stretching, range compression, etc. The critical distribution is the relative frequency of each grey value, the grey value histogram. Image enhancement techniques, while usually not required for automated analysis techniques, have regained a significant interest in current years. Applications such as virtual environments or battlefield simulations require specific enhancement techniques to create 'real life' environments or to process images in near real time, the major focus of these procedures is to enhance imagery data in order to display effectively or record the data for subsequent visual interpretations.

Enhancements are used to make easier visual interpretations and understanding of imagery. The advantage of digital imagery allows manipulating the digital pixel values in an image. Various image enhancement algorithms are applied to remotely sensed data to improve the appearance of an image for human visual analysis or occasionally for subsequent machine analysis. There is no such ideal or best image enhancement because the results are ultimately evaluated by humans, who make subjective judgements whether a given image enhancement is useful. The purpose of the image enhancement is to improve the visual interpretability of an image by increasing the apparent distinction between the features in the scene.

Although radiometric corrections for illumination, atmospheric influences, and sensor characteristics may be done prior to distribution of data to the user, the image may still not be optimized for visual interpretation. Remote sensing devices to cope with levels of target/background energy, which are typically for all conditions, likely to be encountered in routine use. With large variations in spectral response from a diverse range of targets no generic radiometric correction could optimally account for, display the optimum brightness range, and contrast for all targets. Thus, for each application and each imagery, a custom adjustment of the range and distribution of brightness values is usually necessary.

Normally, image enhancement involves techniques for increasing the visual distinctions between features in a scene. The objective is to create new images from the original image data in order to increase the amount of information that can be displayed interactively on a monitor or they can be recorded in a hard copy format either in monochrome or RGB color. Three techniques are categorized as contrast manipulation Gray level threshold, level slicing and contrast stretching, Spatial feature manipulation Spatial filtering, Edge enhancement and Fourier analysis, multi-image manipulation band rationing, differencing, principal components, canonical components, vegetation components, intensity-hue-saturation.

In raw imagery, the useful data often populates only a small portion of the available range of digital values (commonly 8 bits or 256 levels). Contrast enhancement involves changing the original values so that more of the available range is used, thereby increasing the contrast between targets and their backgrounds. The key to understanding contrast enhancements is to understand the concept of an image histogram. A histogram is a graphical representation of the brightness values that comprise an image. The brightness values (i.e. 0-255) are displayed along the x-axis of the graph. The frequency of occurrence of each of these values in the image is shown on the y-axis, through manipulating the range of digital values in an image, graphically represented by its histogram, various enhancements to the data.

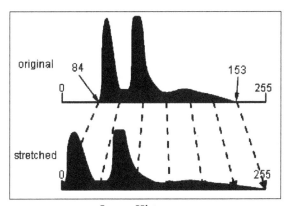

Image Histogram.

There are many different techniques and methods of enhancing contrast and detail in an image. The simplest type of enhancement is a linear contrast stretch. This involves identifying lower and upper bounds from the histogram (usually the minimum and maximum brightness values in the image) and applying a transformation to stretch this range to fill the full range. In the example, the minimum value (occupied by actual data) in the histogram is 84 and the maximum value is 153. These 70 levels occupy less than one-third of the full 256 levels available. A linear stretch uniformly expands this small range to cover the full range of values from 0 to 255.

This enhances the contrast in the image with light toned areas appearing lighter and dark areas appearing darker, making visual interpretation much easier. This illustrates the increase in contrast in an image before (left) and after (right) a linear contrast stretch. A uniform distribution of the input range of values across the full range may not always be an appropriate enhancement, particularly if the input range is not uniformly distributed. In this case, a histogram-equalized stretch may be better. This stretch assigns more display values (range) to the frequently occurring portions of the histogram. In this way, the detail in these areas will be better enhanced relative to those areas of the original histogram where values occur less frequently. In other cases, it may be desirable to enhance the contrast in only a specific portion of the histogram.

For pattern, assume an image of the mouth of a river, and the water portions of the image occupy the digital values from 40 to 76 out of the entire image histogram. Wished to enhance the detail in the water, variations in sediment load, could stretch only that small portion of the histogram represented by the water (40 to 76) to the full grey level range (0 to 255). All pixels below or above these values would be assigned between 0 and 255, respectively, and the detail in these areas would be lost. The detail in the water would be greatly enhanced. Spatial filtering encompasses another set of digital processing functions which are used to enhance the appearance of an image. Spatial filters are designed to highlight or suppress specific features in an image based on their spatial frequency. It refers to the frequency of the variations in tone that appear in an image. "Rough" textured areas of an image, where the changes in tone are abrupt over a small area, have high spatial frequencies, while "smooth" areas with little variation in tone over several pixels, with low spatial frequencies.

A common filtering procedure involves moving a 'window' of a few pixels in dimension (e.g. 3×3, 5×5, etc.) over each pixel in the image, applying a mathematical calculation using the pixel values under that window, and replacing the central pixel with the new value. The window is moved along in both the row and column dimensions one pixel at a time and the calculation is repeated until the entire image has been filtered and a "new" image has been generated. By varying the calculation performed and the weightings of the individual pixels in the filter window, filters can be designed to enhance or suppress different types of features. A low-pass filter is designed to emphasize larger, homogeneous areas of similar tone and reduce the smaller detail in an image. Thus, low-pass filters generally serve to smooth the appearance of an image.

Average and median filters, often used for radar imagery, are examples of low-pass filters. High-pass filters do the opposite and serve to sharpen the appearance of fine detail in an image. One implementation of a high-pass filter first applies a low-pass filter to an image and then subtracts the result from the original, leaving behind only the high spatial frequency information. Directional, or edge detection filters are designed to highlight linear features, such as roads or field boundaries. These filters can also be designed to enhance features which are oriented in specific directions. These filters are useful in applications such as geology, for the detection of linear geologic structures.

Image Enhancement Method

The main aim of the Image Enhancement system is to develop methods that are fast, handle noise efficiently and perform accurate segmentation. For this purpose below methodology uses two stages. The first stage enhances the image in such a way that it improves the segmentation process, while the second step performs the actual segmentation. The working of the enhancement and the segmentation procedure is:

- Step 1: Input Geometric rectified and geometric registration imagery.
- Step 2: Color Conversion.
- Step 3: Image segmentation.
- Step 4: Clustering the Edges.
- Step 5: Image Enhancement Technique:
    - Stage1: Contrast Adjustment.
    - Stage2: Intensity Correction.
    - Stage3: Noise removal.
- Step 6: Enhanced Imagery.

Color Conversion

Most remote sensing systems create arrays of numbers representing an area on the surface of the Earth. The entire array is called an image or scene, and the individual numbers are called pixels (picture elements) such as water body, wetland, forest area etc., the value of the pixel represents a measured quantity such as light intensity over a given range of wavelengths. However, it could also represent a higher-level product such as topography or chlorophyll concentration or almost anything. Some active systems also provide the phase of the reflected radiation so each pixel will contain a complex number. Typical array sizes with optimum pixels and system with multiple channels may require megabytes of storage per scene. Moreover, a satellite can collect 50 of these frames on a single pass so the data sets can be enormous. There are several established color models used in computer graphics, but the most common are the Gray Scale model, RGB (Red-Green-Blue) model, HIS (Hue, Saturation, Intensity) model and CMYK (Cyan-Magenta-Yellow-Black) model.

RGB and L Color Transformation

When Red, Green and Blue light are combined it forms white. As a result to reduce the computational complexity the geo referenced data that exists in RGB color model is converted into a gray scale image. The range of gray scale image from black to white values can be calculated by the equation. Where X is imagery, L is Luminance, R is RED, G is Green and B is Blue.

$$X = L + (0.2989 * R) + (0.5870 * G) + (0.1140 * B)$$

RGB is a color space originated from CRT (or similar) display applications, when it was convenient to describe color as a combination of three colored rays (red, green and blue).

Segmentation using FCM algorithm

Satellite Image Segmentation is one of the most important problems in image preprocessing technique. It consists of constructing a symbolic representation of the imagery that divider an image into non-intersecting regions such that each region is homogeneous and the combination of no two adjacent regions is homogeneous and it can be used for the process of isolating objects of interest from the rest of the scene. Various segmentation algorithms can be found. Starting from the sixties, diverse algorithms have been arising persistently depending upon the applications involved.

Most remote sensing applications image analysis problems need a segmentation section in order to identify the objects or to detect the various boundaries of the imagery and convert it into regions, which are homogeneous according to a given condition, such as surface, color, etc. and assigning labels to every pixel such that pixels with the same label share certain visual characteristics and it's still reflected immature in the field of satellite image processing. The main cause for these vast variations is the image quality while capturing the image and increase in the size of the image and also difficulty in understanding the satellite images by various applications.

The total amount of visual pattern in the image is increased by an overwhelming methodology. These anxieties have increased the use of computers for assisting the processing and analysis of data. The segmentation process in satellite images is considered to be challenging because these images include many textured regions or different background and often subjected to the enlightenment changes or ground truth properties. All these force makes the urgent need in satellite image processing system for rapid and efficient image segmentation model that requires minimum involvement from user. Existing solutions for segmentation of satellite images face three major drawbacks. The representation degradation when supplied with large sized images, degradation of segmentation accuracy due to the quality of the acquired image and speed of segmentation is not meeting the standards of the modern equipments.

This image enhanced considers the use of GIS and remote sensing application of preprocessing segmentation techniques. Preprocessing performs operations on the input imagery to improve the imagery quality and FCM clustering algorithm is to increase the image quality by the segmentation process. It includes Color transformation, intensity correction, method and parameter selection, edge or boundary enhancement and de-noising. Out of these, boundary enhancement, pixel correction and de-noising have more impact on segmented results. ERDAS imaging Segmentation process involves several steps. To input image conversion to particular feature space depends on the clustering techniques which uses two steps:

- Primary step involves the conversion of the input image into L=RGB color value attributes using fuzzy c-means clustering method.

- Secondary step involves the image conversion to feature space with the selected fuzzy c-means clustering method.

The above method paving the way for next segmentation process (input image conversion to feature space of clustering Method). In hard or unsupervised clustering, data is divided into distinct clusters, where each data element belongs to exactly one cluster. In fuzzy clustering, data elements can belong to more than one cluster by using Cluster Center Initialization algorithm, and associated with each element is a set of membership levels. These indicate the strength of the association between that data element and a particular cluster. Fuzzy clustering is a process of assigning these membership levels, and then using them to assign data elements to one or more clusters. The most significant part of this segmentation method is grant of feature value. In the grant of feature value is based on simple idea, that neighboring pixels have approximately same values of lightness and chroma. Then an actual image, noise is corrupting the imagery data or imagery commonly contains of textured segments. Basic segmentation methods based on fuzzy c-means clustering algorithm are working as follows:

Cluster Centers Initialization Method

Required X: dataset, C: no. of Clusters

Procedure ordering-split (X, c)

Compute m for each k $k \in \{1,........,n\}$

Apply to m the ordering function σ

for I = 0 to c do,

$$\ell_i \leftarrow i * \left[n/c \right]$$

end for,

for i=1 to c do

$$S_i \leftarrow \{\ell_{i-1}+1,..........,\ell_i\}$$
$$C_i \leftarrow \sigma^{-1}(S_i)$$
$$V_i \leftarrow \frac{1}{|C_i|} \sum_{j \in c_i} X_j$$

end for

return V

End procedure

Fuzzy C-Mean Algorithm

Procedure Segmentation(Image I, No.of Clusters c, No.of bins q)

Pre-processing the image I

Initialize cluster center v using the ordering-split procedure.

repeat

Update partition matrix U

Update prototypes matrix V

Until is a matrix norm.

Regularize the partition U

Return(U,V)

End procedure

The Fuzzy C-Mean algorithm is described previously, which allots pixels to each class by using fuzzy memberships. Let X = $(x_1, x_2,................,x_N)$ denotes an image with N pixels to be segregated into c clusters, where xi represents multispectral imagery(features) data. The algorithm is an iterative optimization that minimizes the cost function defined as follows:

$$J = \sum_{j=1}^{N}\sum_{i=1}^{c} u_{ij}^m \left\| x_j - v_i \right\|^2$$

where u_{ij} represents the membership of pixel x_j in the ith cluster, v_i is the ith cluster center, $||.||$ is a norm metric, and m is a constant. The parameter m controls the fuzziness of the resulting partition is used in this study.

- Image Smoothing: The edge detection for the given imagery will be done smoothen the image using specific iteration. The specific iteration will be selected for the each image is the tool. If the imagery is noisy, the smoothing process will be applied of the noisy pixel in the process of edge detection.

- Threshold: A thresholding procedure attempts to determine an intensity value, called the threshold, which separates the desired classes. The segmentation is then achieved by grouping all pixels with intensity greater than the threshold as one class, and all other pixels as another class. Thresholding is a simple yet often effective means for obtaining segmentation in images. The limitation of thresholding is that, in its simplest form only two classes are generated and it cannot be applied to multi-channel images. In addition, thresholding does not take into account the spatial characteristics of an image and therefore, are sensitive to noise. For these reasons variations on classical thresholding have been proposed that incorporates information based on local intensities and connectivity.

- Minimal Length: In edge detection process is to accept exact minimum length of the edge. The acceptable length will be measured from the adjacent point of the imagery and if it is less than the acceptable length the segment method will be dropped.

- Region Growing: Region growing is a technique for extracting a region of the image that is connected based on some predefined criteria. These criteria can be based on intensity information and/or edges in the image. Region growing requires a seed point and extracts all pixels connected to the initial seed with the same intensity value. Its primary disadvantage is that it requires manual interaction to obtain the seed point. This problem can be solved by using split and merge algorithms which do not require a seed point. Region growing are sensitive to noise, causing extracted regions to have holes or even become disconnected. Conversely, partial volume effects can cause separate regions to become connected. To help improve these problems, a hemitropic region growing algorithm has been proposed that preserves the topology between an initial region and an extracted region.

Parameter for Locating

In this selection is set to be additional parameters used in edge detection process. There are minimal value difference and variance factor:

- The minimum value is used for neighboring segment between minimal differences.

- The variation factors specify the important role that shows variation in pixel value with in the same segment. This segmented result plays in defining whether expand the segment or not.

Area of interest parameter (AOI) is to use in specify the selected areas of the image to perform the Segmentation process.

Image Parameter Initialization:

- Step 1: Choose a number of clusters in a given image.

- Step 2: Assign randomly to each point coefficients for being in a cluster.

- Step 3: Repeat until convergence criterion is met.
- Step 4: Compute the center of each cluster.
- Step 5: For each point, compute its coefficients of being in the cluster.

The first measures of evaluation of segmentation were subjective, and the ever growing interest in this topic leaded to numerous metrics allowing proper evaluation. In order to objectively measure the quality of the segmentations produced, evaluation measures are considered in the enhancement.

Clustering the Segmented Regions

Clustering algorithms essentially perform the same function as classifier methods without the use of training data and are termed unsupervised methods. In order to compensate for the lack of training data, clustering methods iterate between segmenting the image and characterizing the properties of each class. In short, clustering methods train themselves using the available data. Three commonly used clustering algorithms are k-means and the fuzzy means algorithm. Although clustering algorithms do not require training data, they do require an initial segmentation (or equivalently, initial parameters). Like classifier methods, clustering algorithms do not directly incorporate spatial modeling and can therefore be sensitive to noise and intensity in homogeneities. This lack of spatial modeling, however, provides significant advantages for fast computation. Work on improving the robustness of clustering algorithm to intensity in homogeneities has demonstrated excellent success. Robustness to noise can be incorporated using Markov random field modeling.

Segmentation Results

The satellite images retrieved from various places have been tested in the study area by using ERDAS IMAGING software. The figure gathered from the satellite is given an input to the FCM algorithm where the image undergoes various transformations like Forest, Wetland, Water Body, and River areas are the four different regions selected from the satellite imagery using AOI tools. The below figure (a) Forest, figure (b) Wetland, figure (c) Water Body, figure (d) River are the preferred regions. The satellite imagery does not reveal the clear picture of the selected regions and so the above four figures (a), (b),(c), (d) are distinguished from figure to make the image more visible.

The FCM Algorithm takes as input the above images and segments the images according to the regions with minimum distance. The following images when passed through the FCM algorithm using ERDAS IMAGING software get transformed in to the following images as figure (e), figure (f), figure (g), and figure (h) respectively. The places that are recognized from the scalable imagery using the FCM method generate the segmented results of the selected regions.

Enhanced FCM clustering is a hard and an unsupervised clustering technique which will be applied to image segments to clusters with spectral properties. FCM use the distance between pixels and cluster centers in the spectral domain to compute the membership function. Image pixels are highly correlated, and this spatial information is an important characteristic that can be used to aid their classification.

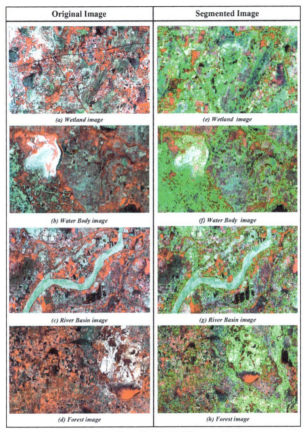

Segmentation Results in Different Regions.

Image Enhancement Techniques

The Enhancement algorithm used for enhancing the input satellite image starts with applying for 2D transformation to obtain four sub-bands, namely, FF, FS, SF and SS. It is known that the FS, SF and SS sub-bands has the edge details while the FF sub-band has the detailed information of an image. The Enhancement method works in two stages as given below:

- Stage 1: Uses Contrast Limited Adaptive Histogram Equalization algorithm to enhance FS, SF and SS sub-bands.

- Stage 2: Performs Intensity correction and removes noise using edge-preserving directional anisotropic diffusion method. Finally, an inverse transformation is performed to obtain the enhanced image.

Contrast Adjustment

The CLAHE algorithm is a special class of adaptive histogram equalization. Adaptive histogram equalization maximizes the contrast throughout an image by adaptively enhancing the contrast of each pixel relative to its local neighborhood. This process produces improved contrast for all levels of contrast (small and large) in the original image. For adaptive histogram equalization to enhance local contrast, histograms are calculated for small regional areas of pixels, producing local histograms. These local histograms are then equalized or remapped from the often narrow range of intensity values indicative of a central pixel and its closest neighbors to the full range of intensity values available in the display. Further, to enhance the edges, a sigmoid function is used,

$$I(X) = \frac{N}{1+e^{-\left(\frac{X-N-\Delta X}{a}\right)}} + \Delta X$$

where M is 255, m = 128 (for 8 bit image), x is the edge pixel, $-127 \Delta \leq x \leq +128$, parameter 'a' refers to the speed of the change around the center. This process is repeated for detailed coefficients. Finally an inverse wavelet transformation is performed to obtain an edge enhanced image.

Intensity Correction

Intensity non-uniformity in satellite images is due to a number of causes during the acquisition of the image data. In principle, they occur due to the non-uniformity of the acquisition devices and relate to artifacts caused by slow, non-anatomic intensity variations. In this paper, an Expectation-Maximization (EM) algorithm is employed to correct the spatial variations of intensity. The Expectation Maximization methods do not make any assumption of the sequences type or texture intensity and therefore can be applied to all kind of image sequences. In general, the EM algorithm consists of two steps: (i) E-Step (or) Expectation Step and (ii) M-Step (or) Maximization step. The algorithm is similar to the K-means procedure in the sense that a set of parameters are re-computed until a desired convergence value is achieved. These two steps are repeated alternatively in an iterative fashion till convergence is reached.

Expectation Maximization Method:

- Step 1: Initialize bias field to mean variance of the image (x) and weight field to the Gaussian value (c) and estimate initial probability as $P = X|C$.
- Step 2: E-Step: Estimate Expected-value of the hidden intensity value for the current value $P_{new}(X_i) = P(X_i|C_i)$.
- Step 3: M-Step: Re-estimate the model parameters by taking the maximum likelihood estimate according to the current estimate of the complete data.

- $P_{old}(X_i) = P_{new}(X_i)$.

- $P_{new}(C_i) = \dfrac{1}{N} \sum_{i=1}^{N} P_{new}(C_i|X_i)$.

- Step 4: If $\dfrac{P_{new}}{P_{old}} \langle 1+ \in$ then convergence is reached, stop process, else go to Step 2.

Noise Removal

After correcting intensity, an enhanced version of anisotropic diffusion is applied to remove speckle noise in a fast and efficient manner. It also called Perona Malik diffusion, is a technique aiming at reducing image noise without removing significant parts of the image content, typically edges, lines or other details that are important for the interpretation of the image. Anisotropic diffusion filter is a frequently used filtering technique in digital images. In spite of its popularity, the established anisotropic diffusion algorithm introduces blocking effects and destroys structural and spatial neighborhood information. Further they are slow in reaching a convergence stage. To solve these problems, the algorithm was combined with an edge-sensitive partial differential equation during new hybrid method of noise removal [YAC02]. The anisotropic filtering in hybrid noise removal simplifies image features to improve image segmentation by smoothening the image in homogeneous area while preserving and enhances the edges. It reduces blocking artifacts by deleting small edges amplified by homomorphic filtering.

Hybrid De-Noised Algorithm:

- Stage 1: Read intensity corrected image.

- Stage 2: Divide into 8 × 8 blocks and for each sub block, perform steps 2a and 2c:

    - Calculate Bayesian shrinkage threshold for each iteration of anisotropic diffusion T = max (r(T)) where r(T) = E(X' − X), where X is the image and X' is the GCD.

    - Perform directional anisotropic diffusion.

    - If convergence reached, then goto step 3, else step 2a.

- Stage 3: Output de-noised image.

This hybrid model is an effective noise removing algorithm but the convergence time still needs to be improved. In this research work, the hybrid noise removing algorithm is further improved by using the numerical characteristics for the instability flow. The concept is to add to the new filter method a non-scalar component which can perform directional filtering of the image along the structures and is therefore named as directional hybrid noise removing. The directional hybrid noise remove while combined with Baye's shrink thresholding, produce faster de-noising operation.

Testing and Results

The Enhancement algorithm is tested with IRS P6 LISS III and LISS IV satellite imagery. There are various Land Use and/or Land Cover Classes appearing on the imagery including river, forest, urban and ext. In the first step, the number of cluster (c), is given by 5 and the fuzziness (f), is given by a value of 2 for purpose of efficient computation. Further, population parameters (Pu) and (pl), are both given by 0.015 to build the stretch model of each cluster. The original image and enhance images using conventional methods and the FCM method. The original image obviously appears that the brightness is dark and the contrast is low. By using the conventional enhancement methods, the gray values with extremely dark or bright are visibly over saturated.

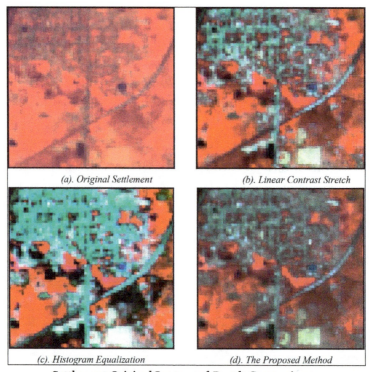

(a). Original Settlement *(b). Linear Contrast Stretch*
(c). Histogram Equalization *(d). The Proposed Method*

Settlement Original Image and Result Comparisons.

As figure above, (d) shows the proposed method provides better visualization in colour and details than other methods in the settlement. Figure (a), (b) and (c) show the comparisons of the enlarge images of forests, and water body respectively. In these areas, the conventional enhancement methods tend to lose the tiny details of the images, while the proposed method could provide more details and better contrast in the image.

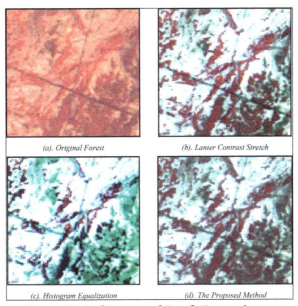

Forest Original Image and Result Comparisons.

As figure above, (d) shows the proposed method provides better visualization in colour and details than other methods in the forests. Figure (a), (b) and (c) show the comparisons of the enlarge images of settlement, and water body respectively. In these areas, the conventional enhancement methods tend to lose the tiny details of the images, while the proposed method could provide more details and better contrast in the image.

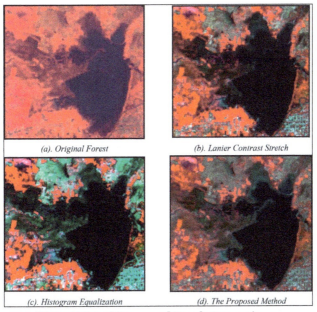

Water Original Image and Result Comparisons.

As figure (d) shows the proposed method provides better visualization in color and details than other methods in the forests. Figure (a), (b) and (c) show the comparisons of the enlarge images of settlement, and water body respectively. In these areas, the conventional enhancement methods tend to lose the tiny details of the images, while the proposed method could provide more details and better contrast in the image.

Image Quality Assessment

As shown in above figures, the proposed method provides significantly better contrast and details for human visual perception than the conventional enhancement methods. However, the visual performance of the contrast enhancement approach is difficult to evaluate and compare with different methods objectively. Hence, a metric index is required to estimate the result. In this study, two indices, entropy and Image Quality Measure, are used to evaluate the results.

Shannon Entropy

Shannon Entropy (or information entropy) is a method to measure the uncertainty of the information. Assume there are n events in the sample space, the probability of each event is pi (i= 1, 2, ..., n), each pi is equal or greater than zero, and the sum of pi is defined to be 1. Therefore, a function H could be defined to measure the uncertainty of the sample space. For image processing, n is given by the number of gray level.

Then the H could be described as equation below. From the values of the entropy, it appears that the information of the image is richer when entropy is higher. Since the test data is multispectral image, the entropy in this study is calculated by averaging all bands. The entropy results are shown in table. Entropy of the image enhanced by the proposed method is 5.071 which are higher than the values of images enhanced by the conventional methods.

$$H = \sum_{i=0}^{L-1} P_i \ln(P_i)$$

where L is represented number of gray level, P_i is represented probability of level i in the histogram.

Image Quality Measure

The proposed a method to measure the quality of natural scene based on human visual system. The algorithm performs as the following steps. First, the image is transformed to power spectrum using Fourier transform. Second, the power spectrum is normalized by brightness and size of the image. Third, a vision filter is used to incorporate with the human visual system model. Moreover, the system needs a noise filter to control the noise of the image and a directional scale factor to treat the images obliquely acquired. Finally, the measure is obtained from the power spectrum weighted by the above processes. It appears that the image quality seems better when Image Quality

Measure index is higher. Table also shows the Image Quality Measure index's of the images enhanced by the proposed method and the conventional methods. The comparison indicates that the image enhanced by the proposed method can obtain higher Image Quality Measure index and accordingly, better quality than the conventional methods.

$$IQM = \frac{1}{S^2} \sum_{\theta=180}^{180} \sum_{\rho=0.01}^{0.5} S(\theta_1)W(\rho)A^2(T_\rho)P(\rho,\theta)$$

where I_2 is represented to image size, $S(\theta_1)$ is represented to directional image scale parameter, $W(\rho)$ is represented to modified Wiener noise filter, $A^2(T_\rho)$ is represented to modulation transfer function of human visual system, $P(\rho, \theta)$ is represented to brightness normalized image power spectrum, ρ, θ is represented to spatial frequency in polar coordinates. Most conventional contrast enhancement algorithms usually fail to provide detailed contrast information in the dark and bright areas of remotely sensed images. This study proposed a fuzzy-based approach to enhance all the contrast and brightness details of the image. The test results indicate that the proposed method could provide better contrast image than the conventional enhancement methods in terms of visual looks and image details. Moreover, two image quality indices are used to evaluate the performance of the enhancement technique.

Table: Image Quality Analysis.

Index	Algorithm Analysis		
	Histogram Equalization	Simple Linear Contrast Stretch	Proposed Method (FCM)
IQM	$4.30*10^{-3}$	$2.92*10^{-3}$	$4.72*10^{-3}$
Entropy	4.049	4.039	5.571

The comparison shows that the proposed method can produce better measurements than the conventional enhancement techniques. However, the stretch method used to enhance each cluster in this study is generated by a linear model with stretch parameters given by experience.

Image Transformation

An important characteristic of digital images, that are captured using multispectral scanners, is that they provide potential to detect the differences between various spectral bands. This allows a further qualitative and/or quantitative assessment, as the difference between two band values can be compared with the spectral signatures of different features on earth. The process of Image transformation, therefore, involves analysis of the data in multiple bands. These bands either may be a single multi-spectral image or from two or more images of an area captured on different dates. It may also be carried out using data from two or more images captured at different spatial resolution.

Thus, in the process of image transformation using multispectral bands, image processing functions can be presented in the same manner as the variables presented in a mathematical expression. For instance, one image band can be added to, or subtracted from, another band on a pixel-by-pixel basis. A new image is obtained using the new pixel values that would highlight particular features or properties of features on the earth. Some important image transformation techniques include Image Arithmetic Operations, Principal Component Transformation, Tasseled Cap Transformation, HSI Color Space Transformation, Fourier Transformation, and Image Fusion.

Image Arithmetic Operations

One of the common techniques of image transformation is the application of arithmetic operations over the image data. This process involves use of arithmetic operators such as addition, subtraction, and multiplication etc. on two or more digital images of the same area that have been properly co-registered. The process may be applied on two images of different spectral bands of a multi-spectral image data or on the individual bands from different images acquired at different time i.e. multi-temporal images.

Image Addition or Image Averaging

Image addition or averaging is generally carried out to reduce overall noise in an image. In this process, new pixel values in the output image may be obtained by averaging the pixel values of corresponding input images. Similarly, temporal averaging of images may be obtained by averaging two images of the same area but acquired on different days. This results in reduction in the speckle of radar images without losing its spatial resolution. Following is an example of image averaging:

18	24	68	20		60	66	24	66		39	45	46	43
70	69	35	30	+	44	67	89	70	=	57	68	62	50
60	78	24	44		34	28	36	68		47	53	30	56
65	55	52	35		64	94	26	58		65	75	39	47

Image Subtraction

The process of image subtraction can also be applied on two images of the same area, which have been co-registered. It is used to detect changes between the two images. The process is more often used to detect changes in images of urban areas in order to know the changes in urban sprawl. Similarly, in order to quantify the forestation or deforestation, the image subtraction process can be used on two or more multi-temporal images of the forest area under study. Following is an example of image subtraction:

98	102	99	95		99	90	92	95		−1	12	7	0
86	9	90	89	−	92	95	95	88	=	−6	4	−5	1
78	92	88	85		78	92	92	80		0	0	−4	5
80	84	86	80		79	84	90	78		1	0	−4	2

In the output image, pixels with brightness value equal to '0' represent an area where no change or very little change has taken place. Other areas where the change has been detected are represented by pixels with non-zero values. The brightness value being higher or lower than '0' exhibit the direction of changes i.e. increase or decrease in a particular feature on the surface of earth.

Indices and Band Rationing

Indices or band rationing is one of the most common arithmetic operations. It is generally applied in interpretation of digital images for various applications including agricultural, forestry, ecological and geological. Typical applications include removal of topographic effects on images and detection of different cover types, including deriving vegetation indices.

A significant difference in DN values, of similar surface features, often results in due to the variation in slope and aspect of the terrain, shadows over the surface, and/or due to the variation in angle of sunlight illumination and the intensity of sunlight. These variations, if not properly corrected for, may lead to error in the interpretation of remote sensing data. It also affects the algorithm to correctly identify the surface materials or the land use, in a remotely sensed image. Use of band ratio can be quite useful under such conditions and therefore it can be applied to reduce the effects of various phenomena.

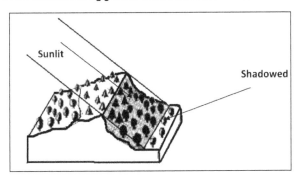

The intensity of illumination that a point on earth's surface receives is a function of the angle of sunlight approaching the slope. If the sun is directly overhead, then the area on the earth's surface receives the most light. Therefore, the amount of light that is reflected back to a sensor is a function of the angle of illumination apart from the properties of the surface material. This property is termed as Topographic Effect. For example, the following figure shows two different land cover types i.e. covered with deciduous and coniferous trees covering both the sunlit and shadowed portions of a terrain.

Table: DN values obtained in a remotely sensed data.

Landcover	Illumination	DN Values		
		Band A	Band B	Ratio (Band A / Band B)
Deciduous	Sunlit	46	51	0.90
	Shadowed	18	20	0.90
Coniferous	Sunlit	30	45	0.67
	Shadowed	10	15	0.67

It may be observed from the above table that there are low DN values in individual bands for the area that is under shadow. In this case, it may be difficult to match the shadowed area with the one in sunlit portion. However, the values of ratio of DN values in different bands, as may be observed from the above table, are almost identical irrespective of the illumination condition of the terrain. Thus, rationing of bands results in an output image in which the problem of variation in DN values of similar type of features on earth's surface due to varying topography is overcome.

Band rationing can also be applied in case the absorption in one band is different from that in another band. This would help in distinguishing several features e.g. stressed vegetation can be distinguished from healthy vegetation; algae can be determined in turbid water; and the parent materials of soils can also be traced. Typical examples also include the one, which is based on the sharp difference in spectral signatures between bare soil and green vegetation. The bare soil exhibits a near linear spectral curve in the region of visible and near infrared (NIR), while the vegetation typically has a low reflectance in red (R) and very high reflectance in the NIR region. Therefore, the ratio of red band to near-infrared band (DN_{Red}/DN_{NIR}) should help distinguish the bare soil from green vegetation. In this case, values of ratio will be large in case of bare soils whereas it will be low for green vegetation. Other indices or band ratios that are commonly used in analysis of remote sensing data include:

- Normalized differential vegetation index (NDVI).
- Soil adjusted vegetation index (SAVI).
- Transformed vegetation index.
- Perpendicular vegetation index.
- Normalized difference snow index.

- Iron oxide index.

Normalized Differential Vegetation Index

One of the common applications of remote sensing technique is monitoring and assessment of vegetation expanding over large areas on the surface of earth. In general, visible and near infrared (NIR) bands are used for this purpose. Various combinations of these bands have been used to develop different kinds of vegetation indices such as a simple vegetation index (VI), rational vegetation index (RVI) and normalized differential vegetation index (NDVI). These indices are expressed as follows:

$$VI = DN_{NIR} - DN_{Red}$$

$$RVI = \frac{DN_{Red}}{DN_{NIR}}$$

$$NDVI = \left[\frac{DN_{NIR} - DN_{Red}}{DN_{NIR} + DN_{Red}}\right]$$

The value of NDVI ranges between −1.0 and +1.0. A positive value of NDVI, i.e. from 0.3 to 0.8, shows that the area is covered with vegetation canopy. A further categorization of vegetation is suggested based on NDVI wherein the value ranging from 0.6 to 0.8 refers to dense vegetation and the range of NDVI from 0.3 to 0.6 refers to stressed vegetation. NDVI values for different features, may however, vary from season to season or from species to species. Clouds, water bodies, and snow covers are generally characterized by negative values of NDVI. Land covers such as soils are represented by the NDVI value equal to or near to zero.

Soil Adjusted Vegetation Index

This is a further modification of normalized differential vegetation index. It may be expressed by the following relationship:

$$SAVI = \left[\frac{DN_{NIR} - DN_{Red}}{DN_{NIR} + DN_{Red} + M}\right](1+M)$$

where, M is a constant that may be determined empirically. The effect of M is such that it minimizes the vegetation index sensitivity to soil. For an intermediate vegetation cover, the value of M is equal to 0.5. It may be noted that if the value of M is equal to 0, then the SAVI is equal to NDVI. Use of SAVI may be preferred to that of NDVI in order to suppress the effect of different soil background on the NDVI.

Transformed Vegetation Index

It is more linearly related to the vegetation biomass as compared to the NDVI. Therefore, it is widely used for estimation of biomass on earth's surface. The equation for transformed vegetation index is as follows:

$$TVI = \sqrt{\left(\frac{DN_{NIR} - DN_{Red}}{DN_{NIR} + DN_{Red}} + 0.5\right)}$$

Perpendicular Vegetation Index

The perpendicular distance of a vector of pixels from the soil line in the feature space plot of NIR and Red band is considered. A feature space plot is a scatter plot of DN brightness values of two bands drawn against each other. It is an indicator of growth of plants. Therefore, red and near infrared bands are used in PVI to determine the perpendicular distance between the vegetation spot on the NIR-Red scatter plot and the soil line. Since vegetation has higher NIR and lower red reflectance than the underlying soil, the vegetation spot will be on the top left corner of the scatter plot. As the density of vegetation increases, the vegetation spot moves further away from the soil line i.e. towards the top left area of the feature space plot.

Values of red band and NIR band pairs representing any water content for the bare soils coincides with the line AB. Point C represents the intermediate vegetative growth with the soil at intermediate water content. The length of the perpendicular drawn from point C to the line AB represents the value of PVI. One of the limitations of the PVI is the assumption that there will be only one type of soil beneath the vegetation. However, this may not be true always, as there may be a mixture of soil types or a mixture of soil and rocks present within a small area.

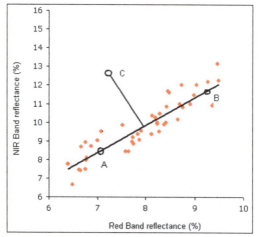

Example of a soil line between red and near infrared reflectance.

Normalized Difference Snow Index

NDSI is also developed on the same principle as that used in development of the normalized differential vegetation index. The reflectance of snow is high in the visible wavelengths i.e. 0.5 μ – 0.7 μ. Whereas, its reflectance is low in the shortwave infrared range of wavelengths, i.e. 1 μ – 4 μ. The NDSI is determined as the difference of reflectance values observed in a visible band (e.g. green band) and a short-wave infrared (SWIR) band, divided by the sum of the reflectance values in the same bands.

$$NDSI = \left[\frac{DN_{Green} - DN_{SWIR}}{DN_{Green} + DN_{SWIR}} \right]$$

Similar to many ratio approaches, the NDSI also tends to reduce the atmospheric effects and the viewing geometry.

Iron Oxide Index

Iron Oxide Index is mathematically expressed as the ratio of the DN values in red band and blue band of an image.

$$\text{Iron Oxide Index} = \left[\frac{DN_{Red}}{DN_{Blue}} \right]$$

This type of index is useful in geological studies involving determination of Iron content in the soil or rock masses.

References

- Properties-of-Aerial-Photography: lpl.arizona.edu, Retrieved 18, March 2020
- Principles-applications-aerial-photography: environmentalscience.org, Retrieved 07, July 2020
- Remote-Sensing-and-GIS-249562471714: igntu.ac.in, Retrieved 28, January 2020

Chapter 3

RADAR and LiDAR

RADAR is Radio Detection And Ranging while LiDAR is Light Detection And Ranging. Radar is an electromagnetic sensor which uses radiowaves to determine the angle, range and velocity of objects. Lidar is a remote sensing method used to measure distances by using light in the form of pulsed laser. The aim of this chapter is to explore the various types of RADAR and LiDAR.

RADAR

The word RADAR is an acronym formed from the expression Radio Detection And Ranging. Common to all radars is the concept of extracting information from a reflected radio signal.

Basic Principle of Operation

- RADAR is fundamentally an electromagnetic sensor used to detect and locate objects.

- Radio waves are radiated out from the radar into free space. Some of the radio waves will be intercepted by reflecting objects (targets).

- The intercepted radio waves that hit the target are reflected back in many different directions. Some of the reflected radio waves (echos) are directed back toward the radar where they are received and amplified.

- With the aid of signal processing a decision is made as to whether or not a target echo signal has been detected. The target location and other information can then be acquired from the echo signal.

Evolution of Radar

The first step towards radar was a simple device to prevent collisions between ships, patented in 1904 by Christian Huelsmeyer. It used crude spark-gap transmitters similar to Marconi's early wireless equipment. Large metallic ships directly ahead of the equipment would increase the spark intensity and cause a bell to ring. Range to target could not be measured, but the principle of targets and echos was established.

The invention of radar is generally attributed to the British as they operated very early radar systems prior to, and during the Second World War. These military equipments were designed to detect and locate enemy ships and aircraft and played a decisive role in the Battle of Britain in 1940. Since then enormous investment has been made in military radar systems and electronic warfare equipment. The steady evolution of military technology has resulted in smaller, more sophisticated and cheaper electronic components that have found subsequent uses for civilian applications. Many of the radar principles that have been proven and refined for military use can be directly applied to commercial radars.

Advances in microprocessor speeds coupled with the development of inexpensive radio components for mobile telephones means it is now possible to produce small sophisticated radar sensors with automatic target detection capabilities at a suitable price for cost-sensitive commercial applications.

Early radar systems used low frequencies so antennas had to be very large to detect distant aircraft - modern microwave radar is much more compact.

Why does Radar use Radio Waves?

Parking sensors on cars use ultrasonic (sound) waves to assist when parking. However, ultrasonic and sound waves only travel at around 330 metres per second, so can only be used over very short ranges. Radio waves are invisible electromagnetic waves that have no mass and travel at the speed of light, approximately 300,000,000 metres per second. The high velocity of electromagnetic waves is ideal for quickly travelling long distances to measure distant objects with minimal delay. There are many different types of electromagnetic waves, such as infrared, X-rays and visible light. Radio waves are used for radar for a number of reasons:

- It is simple and inexpensive to generate radio waves using electronic components.

- Radio waves can pass through fog, rain, mist, snow and smoke.

- Radio waves cannot be confused with infrared energy emitted by fire, heat haze, warm objects, hot gas or the sun.

- Radio waves do not need light to work so radar can operate in total darkness as well as bright sunshine without performance being affected.

- Radio waves are non-ionising so are safe unlike X-rays or gamma rays.

Radio waves have wavelengths between 10,000 km (30Hz frequency) to 1mm (300 GHz frequency). When smaller than 30cm (1 GHz and higher) they are referred to as microwaves. Many radar systems use microwaves because the antennas can be physically smaller as wavelength decreases. Depending on the application the radar designer will select the appropriate operating frequency for best performance.

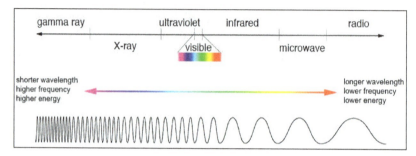

How Radar Measures Range to Target

Radars transmit invisible electromagnetic radio waves that travel at the speed of light, approximately 300 million metres per second. Although this is extremely quick, there will still be a brief delay between the transmission of the original signal and the reception of the echo. The time delay is directly proportional to the range to the target.

Long-range radars use very short pulses and measure the time difference between the original pulse and echo pulse to establish range to target. At shorter ranges a different technique (FMCW) is normally used where the radar constantly transmits but the frequency is modulated so there is a frequency difference between the echo signal and the instantaneous transmitted signal. The radar measures the difference in frequency, which is directly proportional to the range of the target. In both cases, the radar makes a direct measurement of the echo signal to determine range to the object. Compared to optical systems where a large object at long range appears similar to a small object at close range, radar range measurements are not fooled by target size.

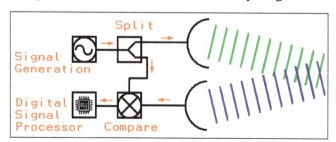

Basic block diagram shows a FMCW radar using two antennas to measure range to target by frequency comparison technique.

How Radar Measures Size of Target

Generally larger objects reflect more radio waves than smaller objects, however the target angle and shape also has an effect. Radar Cross Section (RCS) is the term used to describe the combination of shape and size and is usually expressed in square metres. Targets with higher RCS reflect more radio waves and cause a stronger echo signal to be detected by the radar, so this information can be used to aid in target classification. Although echo strength diminishes with increasing range to target, the radar knows the range from the echo so can compensate for this effect. Typical RCS figures:

- Human 0.5 square metres.
- Car 10 square metres.
- Building 10,000 square metres.

In reality the RCS does vary to some extent based on the target angle, so a building or vehicle that is normal to the radar presents a large flat surface and has a slightly higher RCS than one that is angled so less reflected energy is directed back toward the radar.

Walking humans have an interesting characteristic where the swinging arms and legs caused the RCS to cycle higher and lower in sync with the walking motion. Crawling humans have a lower RCS than walking humans because they have a physically smaller cross section. This can present some difficulties as RCS is similar to small wild animals.

Stealth military aircraft attempt to reduce their RCS by reflecting radio waves away from transmitting radar - note how surfaces are deliberately angled to reflect energy away from radar.

How Radar Measures Speed of Target

Target speed is measured directly by measurement of the Doppler frequency shift. The Doppler Effect is a phenomenon that is regularly experienced even in everyday life. For example when a police siren is heard in the distance the tone changes and rises until the police car drives past and the tone starts to fall again. For radar systems, the Doppler Effect causes moving objects to shift the frequency of reflected radio waves based on the speed of the object. A Doppler shift is seen for objects moving radially, that is, directly toward or away from the radar. Doppler measurement is effective at detecting moving targets and ignoring targets that don't move, which is particularly important for ground surveillance radar where many reflections are seen from stationary targets.

How Radar Measures Direction of Target

Radar antennas typically have a narrow field of view that is scanned across a wider area. When a target is seen the direction in which the antenna is pointing corresponds to the direction of the target. In principle this is like using a telescope to determine the bearing of distant objects. There are many possible antenna methods that can be used, with choice being determined by required size, weight, power and cost. The simplest method is to physically rotate the antenna. When the radar sees the target echo, the direction of the antenna directly corresponds to the direction of the target.

Rotating antennas do have moving parts that can wear, however with clever engineering and use of very small and light materials, the expected lifetime can be extremely long. Some radars have fixed antennas that are steered electronically, a so-called phased array, although this is often much more expensive than simple physical rotation. Another method uses two (or more) antennas to mathematically calculate the angle of arrival by comparing two (or more) echo signals. This method is cheaper than a phased array, but has limitations such as inability to distinguish multiple targets at the same distance and lower sensitivity.

How the Radar Maximum Range is Determined

There are three main factors that determine the maximum range:

- The radar must receive sufficient echo energy to be able to make detection.
- The radar must have a direct line-of-sight to the target.

- Limitations in the receiver circuitry may limit the range that can be measured (instrumented range).

Where the echo energy is the limiting factor, the radar will have different specified ranges for targets of different types, with higher RCS targets being able to be detected further away. Where the limitation is due to the instrumented range the maximum range specification will apply equally to all sized targets. Since microwaves do not bend round corners the target can only be seen if there are no large objects in the way. Mounting the radar higher up allows it to look over objects that are in the way.

Generally it is not possible to increase the transmitted power to attempt to increase the detection range, as the radio regulations are very strict to enforce a maximum power level. This rule is in place to prevent other radio users being negatively affected by interference caused by high power transmissions.

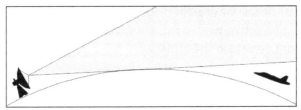

At very long ranges the curvature of the Earth limits the so-called "radar horizon". For short range radars the line-of-sight is usually limited by buildings or obstructions.

How Radars Avoid Interfering with Each Other

When multiple radars of the same type are used in close proximity there must be a method to avoid them causing mutual interference. A common technique is to synchronise equipments to a common clock then slightly offset each radar from its neighbours so they do not transmit on the same frequency at the same time. Another technique is to simply use different operating frequencies that do not overlap at all. Although this doesn't require synchronisation, it severely limits the number of equipments that may be located nearby and is extremely wasteful of the allocated frequency spectrum.

How Clutter Affects Radar Performance

Clutter refers to sources of unwanted echoes generated by objects that reflect radio waves. Clutter is caused by reflections from the ground. Any radar that detects targets on, or close to, the ground will see more clutter than radars that look upwards into the air, especially if the clutter moves.

In the ideal case, the ground would be a very flat concrete expanse with a target located in the middle. Unfortunately this is rare. Often there are fixed objects such as cars, posts, walls and fences that all contribute to the background clutter levels. Fixed clutter can mask the presence of a target by reflecting the radio waves before they can reflect off the target.

Radar signal processing tries to ignore fixed clutter either by filtering objects with no Doppler shift or by comparing the current scan to previous scans to identify fixed objects. Even so, large fixed objects such as tall fences or buildings generate high clutter levels that make it difficult to detect much smaller targets that are next to the clutter due to an effect called scintillation where there are small changes in the echo amplitude of the large object. Consider a large building with RCS of 10,000 square metres exhibiting 0.1% RCS change due to scintillation. This presents a background RCS variation of 10 square metres that is easily enough to swap the RCS of a walking human (1 sq. metre).

Orientating the radar so the building RCS is reduced will improve the situation. Moving clutter, such as long grass, bushes, trees and water is very difficult to mitigate using signal processing. Moving clutter generates a Doppler shift and varies from scan to scan so cannot be distinguished easily from real targets. Radar systems will have a higher detection threshold in areas where there is lots of moving clutter to avoid excessive false alarms. The most effective way to improve the performance is to remove or reduce the moving objects that generate the clutter.

Function of Antennas in Radar Systems

The basic job of the antenna is to emit radio waves when fed with an electrical signal. Antennas are reciprocal, that is, they work just as well in reverse, so the antenna also captures radio waves and emits an electrical signal. Radar may use a single antenna that is shared by transmitter and receiver or may have two antennas, one that transmits, another that receives. Usually pulse radars share a single antenna and FMCW radars use two.

Antennas focus radio waves much like a magnifying glass focuses light. Antenna gain is a measure of the energy increase due to the antenna focussing the radio waves. There is no amplification per se; the gain is simply from focussing more energy into a smaller area. Therefore increasing antenna gain naturally decreases beam width. Antenna gain is influenced by physical dimensions and operating frequency. High frequency microwaves are often used for radar to achieve high antenna gain with physically small antennas.

The main beam width will be halved every time either antenna surface area or the operating frequency is doubled. Effectively this means low frequency radars need physically larger antennas for the same beam width as higher frequency radars that have much smaller antennas. To allow comparison between different radar equipments a common antenna parameter is the -3dB beam width, that is the angle between the points where the antenna power is half that of the peak power measured on boresight.

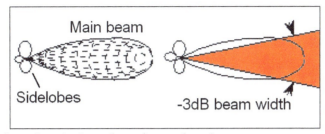

The -3dB point is used as a figure-of-merit to compare antennas.

How Antenna Sidelobes can Affect Radar Performance

All antennas exhibit sidelobes, which are stray signals emitted or received in unwanted directions. It is not possible to produce theoretically perfect antenna that have a single narrow beam with no stray signals at other angles. In reality all antennas are expected to have some imperfections, but the amount of imperfection determines the degradation in system performance.

Sidelobes are characterised relative to the wanted energy of the main boresight beam. High sidelobe levels can cause false detections. For example if a radar sensitivity is set to detect a walking human in the main beam, a sidelobe that has 10% (-10dB) of the energy of the main beam could falsely detect a large target with RCS that is 10x higher than a walking human, for example a large car.

Since the radar always assumes the target is seen in the main beam, the direction to the target will be wrong so would register as a false detection. Clearly if the sidelobe is improved to be 0.1% (-30dB) of main beam energy then the RCS needed for a false detection must be 1000x greater than the human, which would rule out most vehicles, making the radar much less likely to generate false alarms.

Radar manufacturers tend not to specify sidelobes on product datasheets, but estimates can be made from typical antenna pattern graphs (if available) to judge if one system is significantly worse than another.

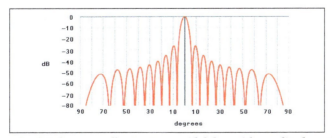

Example plot shows excellent antenna sidelobes with amplitude rapidly reducing away from boresight (0 degrees).

How Antenna Beam Width Affects Radar Operation

Broad antenna beams see more targets (and clutter) at any moment than narrow beams. Narrower main beams aid the radar signal processor in distinguishing one target from another by increasing the angular resolution. Narrow beams also receive fewer reflections from the rest of the environment making it easier to ignore clutter at other bearings.

Radars using angle-of-arrival antennas are unable to use narrow beams, so may struggle compared to systems that rotate a narrow beam. In simplistic terms imagine trying to listen to one person in a crowded noisy room. The angle-of-arrival system is similar to concentrating very hard to try to follow what is being said and to ignore all the other background noise. A narrow antenna is equivalent to blocking out much of the background noise to make it easier to hear the wanted conversation without requiring extraordinary levels of concentration. Narrow beams are most useful in the azimuth (horizontal) plane.

How Vertical Beam Spreading can Improve Performance

A spread beam is an antenna main beam that has been artificially broadened. This differs to a simplistic wide beam by being designed to distribute the radio waves in a specific pattern to achieve good performance at all ranges.

A spread beam cannot be usefully characterised by comparing -3dB points and is not usually symmetrical. This type of shaping is common for radar systems because the range to target has a large effect on the echo power (every time the range doubles the echo power reduces to 1/16 th of what it was), so spread beams distribute the antenna energy to compensate for this effect. Beam spreading is usually in the elevation (vertical) plane only.

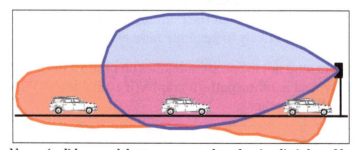

The spread beam (red) has much better coverage than the simplistic broad beam (blue).

Radar Operating Frequency Considerations

Radar users must be aware of the operating frequency and all applicable local and national radio regulations that may prohibit the use of equipment operating at certain frequencies. Broadly speaking, there are two types of frequency: licensed and unlicensed (license exempt).

- License exempt frequencies are the simplest to understand as the equipment may be operated without first obtaining permission. The manufacturer certifies the equipment conforms to license-exempt conditions and the user operates it without further thought.

- Licensed equipment requires the user to first obtain a license from the national radio regulator prior to operating the radar equipment. Licenses may provide benefits for some applications but in almost all cases there are additional fees that must be paid for the privilege.

Types of RADAR

Pulse Doppler Radar

Measuring round trip return timing is fundamental to radar, but it can be difficult to distinguish returns from the target of interest and other objects or background located at similar distances. The use of Doppler processing allows another characteristic of the return to be used – relative velocity. Doppler processing became possible with digital computers and today, nearly all radar systems incorporate Doppler processing.

By measuring the Doppler rate, the radar is able to measure the relative velocity of all objects returning echoes to the radar system – whether planes, vehicles, or ground features. Doppler filtering can be used to discriminate between objects moving at different relative velocities. An example is airborne radar trying to track a moving vehicle on the ground. Since the ground returns will be at the same range as the vehicle, the difference in velocity will be the means of discrimination.

Doppler Effect

The relationship between wavelength and frequency is:

$$\lambda = v/f$$

where:

- f = Wave frequency (Hz or cycles per second).
- λ = Wavelength (meters).
- v = Speed of light (approximately 3×10^8 meters/second).

What happens in a radar system is that the pulse frequency is modified by the process of being reflected by a moving object. Consider the transmission of a sinusoidal wave.

The distance from the crest of each wave to the next is the wavelength, which is inversely proportional to the frequency.

Each successive wave is reflected from the target object of interest. When this object is moving towards the radar system, the next wave crest reflected has a shorter round trip distance to travel, from the radar to the target and back to the radar. This is because the target has moved closer in the interval of time between the previous and current wave crest.

As long as this motion continues, the distance between the arriving wave crests is shorter than the distance between the transmitted wave crests. Since frequency is inversely proportional to wavelength, the frequency of the sinusoidal wave appears to have increased. If the target object is moving away from the radar system, then the opposite happens. Each successive wave crest has a longer round trip distance to travel, so the time between arrival of receive wave crests is lengthened, resulting in a longer (larger) wavelength, and a lower frequency.

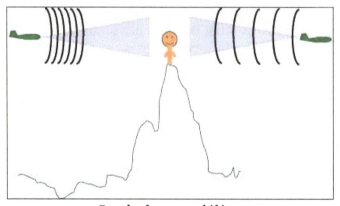

Doppler frequency shifting.

This effect only applies to the motion relative to the radar and the target object. If the object is moving at right angles to the radar there will be no Doppler frequency shift. An example of this would be airborne radar directed at the ground immediately below the aircraft. Assuming level terrain and the aircraft is at a constant altitude, the Doppler shift would be zero, as there is no change in the distance between the plane and ground.

If the radar is ground-based, then all Doppler frequency shifts will be due to the target object motion. If the radar is a vehicle or airborne-based, then the Doppler frequency shifts will be due to the relative motion between the radar and target object.

This can be of great advantage in a radar system. By binning the receive echoes both over range and Doppler frequency offset, target speed as well as range can be determined. Also, this allows easy discrimination between moving objects, such as an aircraft or vehicle, and the back ground clutter, which is generally stationary. For example, imagine there is a radar operating in the X band at 10 GHz (λ = 0.03m or 3cm). The radar is airborne, traveling at 500 mph, is tracking a target ahead moving at 800 mph in the same direction. In this case, the speed differential is −300 mph, or −134 m/s.

Another target is traveling head on toward the airborne radar at 400 mph. This gives a speed differential of 900 mph, or 402 m/s. The Doppler frequency shift can be calculated as follows:

$$f_{Doppler} = 2v_{relative}/\lambda$$

- First target Doppler shift = 2 (−134m/s) / (0.03m) = −8.93 kHz.
- Second target Doppler shift = 2 (402m/s) / (0.03m) = 26.8 kHz.

The receive signal will be offset from 10 GHz by the Doppler frequency. Notice that the Doppler shift is negative when the object is moving away (opening range) from the radar, and is positive when the object is moving towards the radar (closing range).

Pulsed Frequency Spectrum

For this to be of any use, the Doppler shift must be measured. First, the spectral representation of the pulse must be considered. The frequency response of an infinite train of pulses is composed of discrete spectral lines in the envelope of the pulse frequency spectrum. The spectrum repeats at intervals of the PRF.

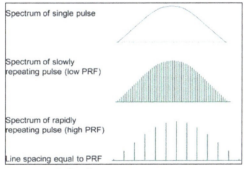

Pulse frequency spectrum.

What is important is that this will impose restrictions on the detectable Doppler frequency shifts. In order to unambiguously identify the Doppler frequency shift, it must be less than the PRF frequency. Doppler frequency shifts greater than this will alias to a lower Doppler frequency. This ambiguity is similar to radar range returns beyond the range of the PRF interval time, as they alias into lower range bins.

$$f_{Doppler} = 2v_{relative}/\lambda$$

Doppler frequency detection is performed by using a bank of narrow digital filters, with overlapping frequency bandwidth (so there are no nulls or frequencies that could go undetected). This is done separately for each range bin. Therefore, at each allowable range, Doppler filtering is applied. Just as the radar looks for peaks from the matched filter detector at every range bin, within every range it will test across the Doppler frequency band to determine the Doppler frequency offset in the receive pulse.

Doppler Ambiguities

Doppler ambiguities can occur if the Doppler range is larger than the PRF. For example, in military airborne radar, the fastest closing rates will be with targets approaching, as both speeds of the radar-bearing aircraft and the target aircraft are summed. This should assume the maximum speed of both aircraft.

The highest opening rates might be when a target is flying away from the radar-bearing aircraft. Here, the radar-bearing aircraft is assumed to be traveling at minimum speed, as well as the target aircraft flying at maximum speed. It is also assumed that the target aircraft is flying a large angle θ from the radar-bearing aircraft flight path, which further reduces the radar-bearing aircraft speed in the direction of the target.

The maximum positive Doppler frequency (fastest closing rate) at 10 GHz/3 cm is:

- Radar–bearing aircraft maximum speed: 1200 mph = 536 m/s.
- Target aircraft maximum speed: 1200 mph = 536 m/s.
- Maximum positive Doppler = 2 (1072m/s) / (0.03m) = 71.5 kHz.

The maximum negative Doppler frequency (fastest opening rate) at 10 GHz/3 cm is:

- Radar-bearing aircraft minimum speed: 300 mph = 134 m/s.
- Effective radar-bearing aircraft minimum speed with θ = 60 degree angle from target track (sin (60) = 0.5): 150 mph = 67 m/s.
- Target aircraft maximum speed: 1200 mph = 536 m/s.
- Maximum negative Doppler = 2 (67–536 m/s) / (0.03m) = 31.3 kHz.

This results in a total Doppler range of 71.5 + 31.3 = 102.8 kHz. Unless the PRF exceeds 102.8 kHz, there will be aliasing of the detected Doppler rates, and the associated ambiguities. If the PRF is assumed at 80 kHz, then Doppler aliasing will occur.

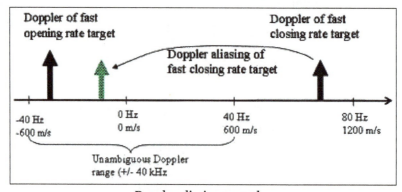

Doppler aliasing example.

Radar Clutter

There are two categories of radar clutter. There is mainlobe clutter and sidelobe clutter. Mainlobe clutter occurs when there are undesirable returns in the mainlobe or within the radar beamwidth. This usually occurs when the mainlobe intersects the ground. This can occur because the radar is aimed downward (negative elevation), there is higher ground such as mountains in the radar path, or even if the radar beam is aimed level and as the beam spreads with distance hits intersects the ground. Because the area of ground in the radar beam is often large, the ground return can be much larger than target returns.

Sidelobe clutter is unwanted returns that are coming from a direction outside the mainlobe. Sidelobe clutter is usually attenuated by 50 dB or more, due to the antenna directional selectivity or directional radiation pattern. A very common source of sidelobe clutter is ground return. When radar is pointed toward the horizon, there is a very large area of ground area covered by the sidelobes in the negative elevation region. The large reflective area covered by the sidelobe can cause significant sidelobe returns despite the antenna attenuation.

Different types of terrain will have a different "reflectivity", which is a measure of how much radar energy is reflected back. This also depends on the angle of the radar energy relative to the ground surface. Some surfaces, like smooth water, reflect most of the radar energy away from the radar transmitter, particularly at shallow angles. A desert would reflect more of the energy back to the radar, while wooded terrain would reflect even more. Man-made surfaces, such as in urban areas tend to reflect the most energy back to the radar system.

Often targets are moving, and Doppler processing is an effective method to distinguish the target from the background clutter of the ground. However, the Doppler frequency of the ground will can be non-zero if the radar is in motion. Different points on the ground will have different Doppler returns, depending on how far ahead or behind the radar-bearing aircraft that a particular patch of ground is located. Doppler sidelobe clutter can be present over a wide range of Doppler frequencies.

Mainlobe clutter is more likely to be concentrated at a specific frequency, since the mainlobe is far more concentrated (typically 3 to 6 degrees of beam width), so the patch of ground illuminated is likely to be far smaller and all the returns at or near the same relative velocity. A simple example can help illustrate how the radar can combine range and Doppler returns to obtain a more complete picture of the target environment.

Figure illustrates unambiguous range and Doppler returns. This assumes the PRF is low enough to receive all the returns in a single PRF interval and the PRF is high enough to include all Doppler return frequencies.

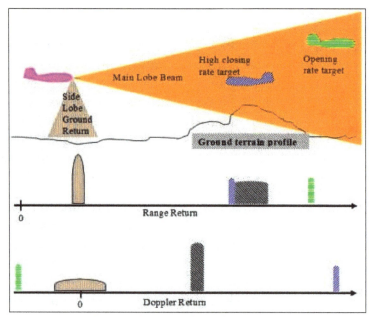

Interpreting Doppler radar returns.

The ground return comes though the antenna sidelobe, known as sidelobe clutter. The reason ground return is often high is due to the amount of reflective area at close range, which results in a strong return despite the sidelobe attenuation of the antenna. The ground return will be at short range, essentially the altitude of the aircraft. In the mainlobe, the range return of the mountains and closing target are close together, due to similar ranges. It is easy to see how if just using the range return, it is easy for a target return to be lost in high terrain returns, known as mainlobe clutter.

The Doppler return gives a different view. The ground return is centered on 0 Hz. The ground slightly ahead of the radar-bearing plane is at slightly positive relative velocity, and the ground behind the plane is at slightly negative relative velocity. As the horizontal distance from the radar-bearing plane increases, the ground return weakens due to increased range.

The Doppler return from mountain terrain is now very distinct from the nearby closing aircraft target. The mountain terrain is moving at a relative velocity equal to the radar-bearing plane's velocity. The closing aircraft relative velocity is the sum of both aircrafts velocity, which is much higher, producing a Doppler return with a high velocity. The other target aircraft, which is slowly opening the range with radar-bearing aircraft, is represented as a negative Doppler frequency return.

PRF Tradeoffs

Different PRF frequencies have different advantages and disadvantages. Low PRF operation is generally used for maximum range detection. It usually requires a high power transmit power, in order to receive returns of sufficient power for detection at a long range. To get the highest power, long transmit pulses are sent, and correspondingly long matched filter processing (or pulse compression) is used. This mode is useful for precise range determination. Strong sidelobe returns can often be determined by their relatively close ranges (ground area near radar system) and filtered out.

Disadvantages are that Doppler processing is relatively ineffective due to so many overlapping Doppler frequency ranges. This limits the ability to detect moving objects in the presence of heavy background clutter, such as moving objects on the ground.

High PRF operation spreads out the frequency spectrum of the receive pulse, allowing a full Doppler spectrum without aliasing or ambiguous Doppler measurements. A high PRF can be used to determine Doppler frequency and therefore relative velocity for all targets. It can also be used when a moving object of interest is obscured by a stationary mass, such as the ground or a mountain, in the radar return. The unambiguous Doppler measurements will make a moving target stand out from a stationary background. This is called mainlobe clutter rejection or filtering. Another benefit is that since more pulses are transmitted in a given interval of time, higher average transmits power levels can be achieved. This can help improve the detection range of a radar system in high PRF mode.

Medium PRF operation is a compromise. Both range and Doppler measurements are ambiguous, but each will not be aliased or folded as severely as the more extreme low or high PRF modes. This can provide a good overall capability for detecting both range and moving targets. However, the folding of the ambiguous regions can also bring a lot of clutter into both range and Doppler measurements. Small shifts in PRFs can be used to resolve ambiguities, but if there is too much clutter, the signals may be undetectable or obscured in both range and Doppler.

FM Ranging

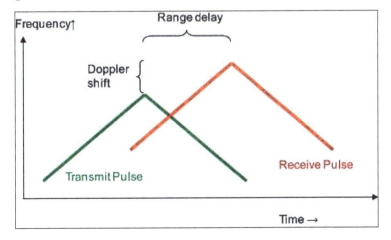

One solution is to use the high PRF mode to identify moving targets, especially fast moving targets, and then switch to a low PRF operation to determine range. Another alternative is to use a technique called FM ranging. In this mode, the transmit duty cycle becomes 100% and the radar transmits and receives continuously.

The transmission is a continuously increasing frequency signal, and then at the maximum frequency, abruptly begins to continuously decrease in frequency until it reaches the minimum frequency. This cycle then repeats. The frequency over time looks like a "saw tooth wave". The receiver can operate while during transmit operation, as the receiver is detecting time delayed versions of the transmit signal, which is at a different frequency than current transmit operation. Therefore, the receiver is not desensitized by the transmitter's high power at the received signal frequency.

Through Doppler detection of what frequency is received, and knowing the transmitter frequency ramp timing, can be used to determine round-trip delay time, and therefore range. And the receive frequency "saw tooth" will be offset by the Doppler frequency. On a rapidly closing target, the receive frequencies will be all offset by a positive $f_{Doppler}$, which can be measured by the receiver once the peak receive frequency is detected.

Moving Target Indicator RADAR

If the Radar is used for detecting the movable target, then the Radar should receive only the echo signal due to that movable target. This echo signal is the desired one. However, in practical applications, Radar receives the echo signals due to stationary objects in addition to the echo signal due to that movable target.

The echo signals due to stationary objects (places) such as land and sea are called clutters because these are unwanted signals. Therefore, we have to choose the Radar in such a way that it considers only the echo signal due to movable target but not the clutters. For this purpose, Radar uses the principle of Doppler Effect for distinguishing the non-stationary targets from stationary objects. This type of Radar is called Moving Target Indicator Radar or simply, MTI Radar. According to Doppler Effect, the frequency of the received signal will increase if the target is moving towards the direction of Radar. Similarly, the frequency of the received signal will decrease if the target is moving away from the Radar.

Types of MTI Radars

We can classify the MTI Radars into the following two types based on the type of transmitter that has been used: MTI Radar with Power Amplifier Transmitter and MTI Radar with Power Oscillator Transmitter.

MTI Radar with Power Amplifier Transmitter

MTI Radar uses single Antenna for both transmission and reception of signals with the help of Duplexer. The block diagram of MTI Radar with power amplifier transmitter is shown in the figure.

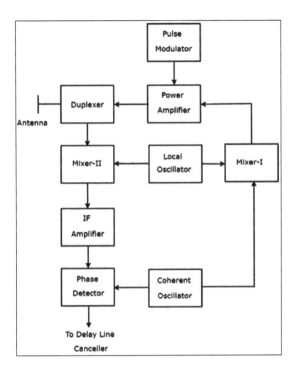

The function of each block of MTI Radar with power amplifier transmitter is mentioned below:

- Pulse Modulator: It produces a pulse modulated signal and it is applied to Power Amplifier.

- Power Amplifier: It amplifies the power levels of the pulse modulated signal.

- Local Oscillator: It produces a signal having stable frequency f_l. Hence, it is also called stable Local Oscillator: The output of Local Oscillator is applied to both Mixer-I and Mixer-II.

- Coherent Oscillator: It produces a signal having an Intermediate Frequency, f_c This signal is used as the reference signal. The output of Coherent Oscillator is applied to both Mixer-I and Phase Detector.

- Mixer-I: Mixer can produce either sum or difference of the frequencies that are applied to it. The signals having frequencies of f_l and f_c are applied to Mixer-I. Here, the Mixer-I is used for producing the output, which is having the frequency $f_l + f_c$.

- Duplexer: It is a microwave switch, which connects the Antenna to either the transmitter section or the receiver section based on the requirement. Antenna transmits the signal having frequency $f_l + f_c$ when the duplexer connects the Antenna to power amplifier. Similarly, Antenna receives the signal having frequency of $f_l + f_c \pm f_d$ when the duplexer connects the Antenna to Mixer-II.

- Mixer-II: Mixer can produce either sum or difference of the frequencies that are applied to it. The signals having frequencies $f_l + f_c \pm f_d$ and f_l are applied to Mixer-II. Here, the Mixer-II is used for producing the output, which is having the frequency $f_c \pm f_d$.

- IF Amplifier: IF amplifier amplifies the Intermediate Frequency (IF) signal. The IF amplifier shown in the figure amplifies the signal having frequency $f_c + f_d$. This amplified signal is applied as an input to Phase detector.

- Phase Detector: It is used to produce the output signal having frequency f_d from the applied two input signals, which are having the frequencies of $f_c + f_d$ and f_c. The output of phase detector can be connected to Delay line canceller.

MTI Radar with Power Oscillator Transmitter

The block diagram of MTI Radar with power oscillator transmitter looks similar to the block diagram of MTI Radar with power amplifier transmitter. The blocks corresponding to the receiver section will be same in both the block diagrams. Whereas, the blocks corresponding to the transmitter section may differ in both the block diagrams. The block diagram of MTI Radar with power oscillator transmitter is shown in the following figure:

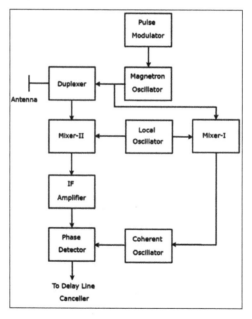

As shown in the figure, MTI Radar uses the single Antenna for both transmission and reception of signals with the help of Duplexer. The operation of MTI Radar with power oscillator transmitter is mentioned below:

- The output of Magnetron Oscillator and the output of Local Oscillator are applied to Mixer-I. This will further produce an IF signal, the phase of which is directly related to the phase of the transmitted signal.

- The output of Mixer-I is applied to the Coherent Oscillator. Therefore, the phase of Coherent Oscillator output will be locked to the phase of IF signal. This means, the phase of Coherent Oscillator output will also directly relate to the phase of the transmitted signal.

- So, the output of Coherent Oscillator can be used as reference signal for comparing the received echo signal with the corresponding transmitted signal using phase detector.

Continuous Wave RADAR

Continuous Wave Radar (CW radar) sets transmit a high-frequency signal continuously. The echo signal is received and processed permanently. One has to resolve two problems with this principle:

- Prevent a direct connection of the transmitted energy into the receiver (feedback connection).

- Assign the received echoes to a time system to be able to do run time measurements.

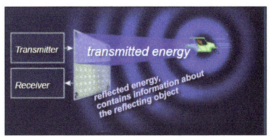

The continuous wave radar methods often uses separate transmit and receive antennas. These are constructed on a double-sided printed circuit board.

A direct connection of the transmitted energy into the receiver can be prevented by:

- The spatial separation of the transmitting antenna and the receiving antenna, e.g. the aim is illuminated by a strong transmitter and the receiver is located in the missile flying directly towards the aim.

- Frequency dependent separation by the Doppler-frequency during the measurement of speeds.

A run time measurement isn't necessary for speed gauges, the actual range of the delinquent car doesn't have a consequence. If you need range information, then the time measurement can be realized by frequency modulation or phase keying of the transmitted power.

A CW-radar transmitting an unmodulated power can measure the speed only by using the Doppler- effect. It cannot measure a range and it cannot differentiate between two or more reflecting objects. If an echo signal is received, this is initially only proof that there is an obstacle in the direction of propagation of the electromagnetic waves.

The properties of the obstacle can be inferred from certain properties of the echo signal. For example, the strength of the echo signal depends on the size of the obstacle. Likewise, the strength of the echo signal is a sign of whether this obstacle is far away or near the radar. (Unfortunately, no measurement result is possible from this context, since the strength of the echo signal depends on too many factors.) A change in the frequency spectrum, on the other hand, is a safer feature for certain properties. Thus, harmonics of the transmission frequency can also occur during a reflection on certain materials. This is specifically exploited in so-called "harmonic radar" in order to use these materials, which are incorporated into protective clothing, for example, to find people buried under the masses of snow in avalanche regions. However, the most commonly used changes in the spectrum are caused by the Doppler Effect.

Doppler Radar

Unmodulated continuous wave radar emits a constant frequency with constant amplitude. The received echo signal either has exactly the same frequency, or the echo signal is shifted by the Doppler frequency (with a reflector moving at a radial velocity). CW radars that specialize in measuring this Doppler frequency are called Doppler radars.

A runtime measurement is not necessary at all with a Doppler radar for speed measurement, since no distance determination is carried out. If a runtime measurement is to be carried out, then a time reference of the received echo to the transmitted signal can be established by modulating the transmitted signal. This modulation, i.e. the actual time at which the transmitted signal changes in frequency or amplitude, can be registered in the receiver after the delay time and thus makes time measurement possible. Such modulation, however, results in other radar classes, which subsequently use completely different measurement principles (for example, frequency modulation: FMCW radar). Amplitude modulation at 100% modulation is also conceivable and would lead to a pulse radar. A radar that emits an unmodulated oscillation can only detect the speed of an object via the Doppler effect. It is not possible to determine distances or distinguish different targets in the same direction.

Function

The continuous wave radar evaluates the phase difference φ between the transmitted signal and the received signal. The magnitude of this phase difference is the ratio of the distance traveled by the electromagnetic wave to the wavelength of the transmitted signal, multiplied by the degree division of the full circle ($2 \cdot \pi$). The magnitude of this phase difference is the ratio of the wavelength of the transmitted signal to the distance traveled. If the distance to the reflector does not change, then it is constant and is calculated according to:

$$\varphi = -2\pi \frac{2r}{\lambda}$$

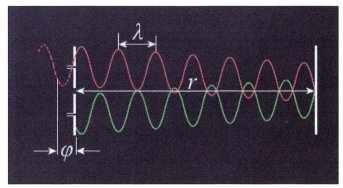

Phase difference.

where,

- φ = Phase difference.
- r = Distance of the reflector to the antenna.
- λ = Wavelength of the transmitted signal.

The factor 2 to the distance r means that the signal has to pass through this distance twice (round trip). The minus sign is generated because a phase jump of 180° occurs during reflection. A direct calculation of a distance from this phase difference is not possible. For example, it would only be possible if the distance were between 0 and 2π $(\triangleq \varphi < 360°)$. From this distance, ambiguities arise due to the periodicity of the sine wave.

If the distance to the reflector is not constant but changes, for example, with a relatively constant speed to the transmitting antenna, the phase difference also changes as a function of time:

$$\varphi(t) = -4\pi \frac{r(t)}{\lambda}$$

A time-dependent and constant change in the phase difference between two sinusoidal signals over the measurement period corresponds to a sinusoidal signal curve again. This can also be measured as a frequency: the Doppler frequency. In most cases, this is in the audio frequency range. At a constant transmission frequency, this Doppler frequency is proportional to the radial speed.

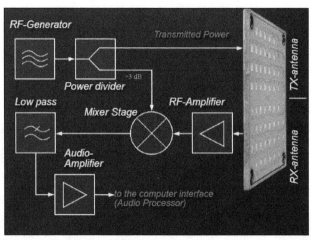

Block diagram of a simple radar transceiver using direct down conversion.

Direct Conversion Receiver

A Doppler radar for speed measurements is very simple. The entire circuit of the transmitter and receiver can be manufactured with semiconductor components on a substrate as an integrated component. This component is usually called a transceiver (a portmanteau of the words transmitter and receiver). In many cases, this transceiver is already equipped with the required antennas. Usually, these are patch antennas realized on a double-sided printed circuit board or (with larger bandwidths) horn radiators.

With a receiver using direct conversion (or referred to a homodyne receiver), the echo signal is not converted into an intermediate frequency, but the high-frequency generated in the transmitter is also used directly for down converting. The output signal of the mixing stage is then in the baseband, i.e. it's free of any carrier frequency. The mixers used usually require a local oscillator power of approximately 7 dBm in order to be able to down convert the echo signal. Thus, the power of the HF generator is also set at 10 dBm. Since the power splitter has a minimum attenuation of -3 dB, the transmission power of at least 6 dBm is specified for the entire module. Although the output signal is now in the baseband, this output is still often referred to as "IF", which suggests an intermediate frequency. However, the Doppler frequency is usually in the audible range. If this DC voltage is not blocked by coupling capacitors as high-pass filters then strong fixed target echoes occur at this output as DC voltage. Usually, such a circuit measure is also carried out in order to prevent signals that are generated by crosstalk from the transmitting antenna to the receiving antenna.

Example: The maximum possible radial velocity v shall be calculated for a Doppler radar in K-Band, ($\lambda \approx 12\ mm$) used as motion detector. How fast can a reflector be moved to process the echo signal with a stereo audio processor of a standard sound card? (f_{cut} = 18 kHz = maximum f_D).

RADAR and LiDAR

In radar, the Doppler frequency is calculated according to:

$$f_D = \frac{2 \cdot v}{\lambda}$$

- f_D = Doppler frequency [Hz].
- λ = The wavelength of the transmitted frequency [m].
- v = Radial velocity [m/s].

This equation was converted according to v and the specified values were inserted:

$$v = \frac{\lambda \cdot f_D}{2} = \frac{12\ mm \cdot 18\ kHz}{2} = 108\ m/s \approx 380\ km/h$$

With this configuration, a maximum of 380 km/h can be measured, which includes most cases of application for a simple motion detector.

Superheterodyne

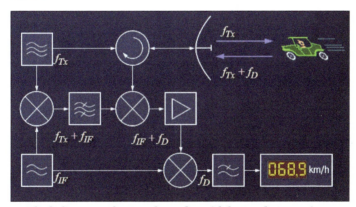

Block diagram of a Doppler radar with heterodyne receiver.

By direct mixing, the sensitivity is limited. Thus, the flicker noise of the mixer is given along with the output signal, i.e., the Doppler frequency is superimposed with a random distribution of low-frequency noise. Very weak signals and low Doppler frequencies cannot be evaluated so often.

A significant improvement in sensitivity may provide at this point a superheterodyne receiver. The echo signals are converted first in a region which is well above the flicker noise. This flicker noise of the first mixer stage cannot pass through the bandpass filter of the intermediate frequency amplifier. Simultaneously, the echo signal is amplified by about 30...40 dB. Only in the second mixer, the echo signal is converted into the baseband. Since the amplified echo signal is now much larger than the flicker noise of the second mixing stage, this noise of the second mixer can be ignored.

In this example, only a single antenna for transmission and reception is used. The separation of the transmission energy of the received energy is performed with a circulator. The local oscillator frequency for the superheterodyne receiver is generated here by an upward mixing followed by a narrow-band filter. As an evaluation of the velocity, a counter is used herein. So that only a single reflective object may be displayed. (Which is usually the one with the highest amplitude). If the radar observes a plurality of moving reflectors, then the overlapping Doppler frequencies need to be selected by a bank of filters, or a tunable filter. Nevertheless, several Doppler frequencies are possible to be measured, there is no way to attribute the simultaneously measured values to a particular object without any doubt.

Calculation of Radar Range

In general, basic radar equation can also be used for continuous wave radar, since it is independent of the modulation type.

$$R_{max} = \sqrt[4]{\frac{P_S \cdot G^2 \cdot \lambda^2 \cdot \sigma}{P_{E_{min}} \cdot (4\pi)^3 \cdot L_{ges}}}$$

However, it must be borne in mind that the losses contained in the term L_{ges} can also contain gains, for example through coherent integration.

Physicists would now point out that the decisive factor for the range of radar is not the transmitting power stated in the formula but the transmitted energy. This could previously be neglected in the derivation of the equation because the duration of the transmission pulse was assumed to be equal to the duration of the echo signal.

If the time duration of the demodulated echo signal differs from the time duration of the transmitted signal, these times must be put into relation and are multiplied as gain with the remaining values of the expression under the fourth root.

$$R_{max} = \sqrt[4]{\frac{P_S \cdot T \cdot G^2 \cdot \lambda^2 \cdot \sigma}{P_{E_{min}} \cdot \tau_c \cdot (4\pi)^3 \cdot L_{ges}}}$$

In radar, with intrapulse modulation, this gain is called pulse compression ratio (PCR) and depends on the transmitted bandwidth. This is illustrated by the fact that this transmitted modulation pattern can hardly be reproduced by random noise pulses so that the pulse compression filter can also detect targets far below the noise level.

A similar calculation can also be made for continuous wave radar. Here, the integration gain can be calculated either with the dwell time in relation to the filter reaction time (CW-radar) or with the bandwidth (FMCW-radar).

Applications of Unmodulated Continuous Wave Radar

1. Traffic control radar (Speed gauges): Speed gauges are very specialized CW-radars. A speed gauge uses the Doppler-frequency for measurement of the speed. Since the value of the Doppler-frequency depends on the wavelength, these radar sets use a very high frequency band. The figure shows the speed gauge.

2. Doppler radar motion sensor: Simple and inexpensive Doppler radar sensors with circuitry can trigger switching functions such as alarms or simply be used as a door opener or switch for lighting.

3. Motion monitoring: If the output of the mixer stage is DC coupled in Figure (i.e. in the mixing stage no inductive transformer or coupling capacitors are used) and the subsequent amplifiers are also all DC coupled, then this non-modulated continuous wave radar can monitors also distances to fixed targets with an accuracy in the order of about $\lambda/16$. Here no Doppler frequency is measured, it is compared the phase angle between the transmitted signal and the received signal. The way from the radar to the reflector and the way back is a multiple of the used wavelength. If this distance changes only by fractions of a millimeter, then the phase angle between the two signals also changes. The measuring range is ambiguous: how many full wavelengths must be additionally added to the measured fraction, this cannot be determined. The radar can monitor only changing to a previous value.

By this measurement method, for example, non-contact monitoring of heart rate and respiratory activity of intensive care patients can be performed. The radar is aligned on the chest of the patient and monitors the distance to an accuracy of fractions of a millimeter. The changes of the phase angle between the transmitted signal and the received signal are displayed on an oscilloscope as a function of time. A connected to the radar computer counts the periodic changes and outputs the heart rate of the patient numerically. If no more changes are registered, an alarm is triggered.

4. The absence of the minimum measuring range typical for pulse radar makes it possible to use this radar system design as a radio proximity fuse for missiles and artillery projectiles. The amplitude of the audible signal increases with the approach to the

target, while the Doppler frequency decreases shortly before passing. Once the passband or low-pass of a filter has been reached, the radar proximity fuse triggers the warhead.

Applications of RADAR

- Military Applications: It has three major applications in the Military:
  - In air defense, it is used for target detection, target recognition, and weapon control (directing the weapon to the tracked targets).
  - In a missile system to guide the weapon.
  - Identifying enemy locations on the map.
- Air Traffic Control: It has three major applications in Air Traffic control:
  - To control air traffic near airports. The Air Surveillance RADAR is used to detect and display the aircraft's position in the airport terminals.
  - To guide the aircraft to land in bad weather using Precision Approach RADAR.
  - To scan the airport surface for aircraft and ground vehicle positions.
- Remote Sensing: It can be used for observing whether or observing planetary positions and monitoring sea ice to ensure a smooth route for ships.
- Ground Traffic Control: It can also be used by traffic police to determine the speed of the vehicle, controlling the movement of vehicles by giving warnings about the presence of other vehicles or any other obstacles behind them.
- Space: It has four major applications:
  - To guide the space vehicle for a safe landing on the moon.
  - To observe the planetary systems.
  - To detect and track satellites.
  - To monitor the meteors.

LiDAR

LiDAR, or light detection and ranging, is a popular remote sensing method used for measuring the exact distance of an object on the earth's surface. Even though it was first used in the 1960s when laser scanners were mounted to aeroplanes, LiDAR didn't get the popularity it deserved until twenty years later. It was only during the 1980s after the introduction of GPS that it became a popular method for calculating accurate geospatial measurements.

LiDAR Technology

LiDAR uses a pulsed laser to calculate an object's variable distances from the earth surface. These light pulses — put together with the information collected by the airborne system — generate accurate 3D information about the earth surface and the target object. There are three primary components of a LiDAR instrument — the scanner, laser and GPS receiver. Other elements that play a vital role in the data collection and analysis are the photodetector and optics. Most government and private organizations use helicopters, drones and airplanes for acquiring LiDAR data.

Working of LiDAR

The working principle of Light Detection and Ranging system is really quite simple. A LIDAR sensor mounted on an aircraft or helicopter. It generates Laser pulse train, which sent to the surface/target to measure the time and it takes to return to its source. The actual calculation for measuring how far a returning light photon has traveled to and from an object is calculated by:

Distance = (Speed of Light x Time of Flight)/2

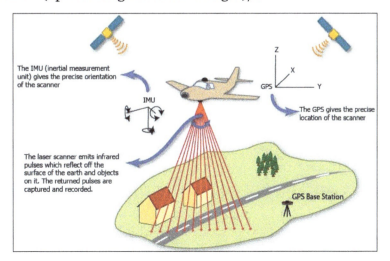

Accurate distances are then calculated to the points on the ground and elevations can be determined along with the ground surface buildings, roads, and vegetation can be recorded. These elevations are combined with digital aerial photography to produce a digital elevation model of the earth.

The laser instrument fires rapid pulses of laser light at a surface, some at up to 150,000 pulses per second. A sensor on the instrument measures the amount of time takes for each pulse to reflect back. Light moves at a constant and known speed so the LiDAR instrument can calculate the distance between itself and the target with high accuracy. By repeating this in quick progression the instrument builds up a complex 'map' of the surface it is measuring.

With airborne Light Detection and Ranging, other data must be collected to ensure accuracy. As the sensor is moving height, location and orientation of the instrument must be included to determine the position of the laser pulse at the time of sending and the time of return. This extra information is crucial to the data's integrity. With ground-based Light Detection and Ranging a single GPS location can be added at each location where the instrument is set up.

LiDAR System Types

Based on the Platform:

- Ground-based LiDAR,
- Airborne LiDAR,
- Spaceborne LiDAR.

Ground-based LiDAR.

Airborne LiDAR.

Spaceborne LiDAR.

Based on Physical Process:

- Rangefinder LiDAR,
- DIAL LiDAR,
- Doppler LiDAR.

Based on Scattering Process:

- Mie,
- Rayleigh,
- Raman,
- Fluorescence.

Main Components of LiDAR System

Most Light Detection and Ranging systems use four main components:

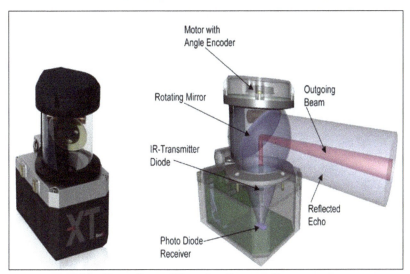

Light Detection and Ranging Systems Components.

Lasers

The Lasers are categorized by their wavelength. Airborne Light Detection and Ranging systems use 1064nm diode-pumped Nd: YAG lasers whereas Bathymetric systems use 532nm double diode-pumped Nd: YAG lasers which penetrate into the water with less attenuation than the airborne system (1064nm). Better resolution can be attained with shorter pulses provided the receiver detector and electronics have sufficient bandwidth to manage the increased data flow.

Scanners and Optics

The speed at which images can be developed is affected by the speed at which it can be scanned into the system. A variety of scanning methods is available for different resolutions such as azimuth and elevation, dual axis scanner, dual oscillating plane mirrors, and polygonal mirrors. The type of optic determines the range and resolution that can be detected by a system.

Photodetector and Receiver Electronics

The photodetector is a device that reads and records the backscattered signal to the system. There are two main types of photodetector technologies, solid state detectors, such as silicon avalanche photodiodes and photomultipliers.

Navigation and Positioning Systems/GPS

When a Light Detection and Ranging sensor is mounted on an aeroplane satellite or automobiles, it is necessary to determine the absolute position and the orientation of the sensor to maintain useable data. Global Positioning Systems (GPS) provide accurate geographical information regarding the position of the sensor and an Inertial Measurement Unit (IMU) records the accurate orientation of the sensor at that location. These two devices provide the method for translating sensor data into static points for use in a variety of systems.

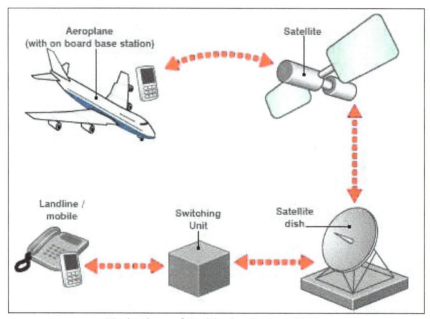

Navigation and Positioning Systems/GPS.

LiDAR Data Processing

The Light Detection and Ranging mechanism just collect elevation data and along with the data of Inertial Measuring Unit is placed with the aircraft and a GPS unit. With the help of these systems the Light Detection and Ranging sensor collect data points, the location of the data is recorded along with the GPS sensor. Data is required to process the return time for each pulse scattered back to the sensor and calculate the variable distances from the sensor, or changes in land cover surfaces. After the survey, the data are downloaded and processed using specially designed computer software (LiDAR-point Cloud Data Processing Software). The final output is accurate, geographically registered longitude (X), latitude (Y), and elevation (Z) for every data point. The LiDAR mapping data are composed of elevation measurements of the surface and are attained through aerial topographic surveys. The file format used to capture and store LiDAR data is a simple text file. By using elevation points data may be used to create detailed topographic maps. With these data points even they also allow the generation of a digital elevation model of the ground surface.

Types of LiDAR

LiDAR has various applications in surveying, sensors, drones and laser scanning. So people are fairly acquainted with this technology, which is among the major emerging technologies driving the age of automation. But there are many different types of LiDARs based on their functionality and inherent characteristics. Let's have a look at the classification of LiDARs and how different types of LiDARs have different uses.

Based on Functionality

Airborne

As the name suggests, Airborne LiDARs are mounted on top of a helicopter or drone. The light is first emitted towards the ground and then it moves towards the sensor. Airborne LiDAR is further classified into topographic and bathymetric.

- Topographic LiDAR: It is used mainly in monitoring and mapping topography of a region. So it has its applications in geomorphology, urban planning, landscape ecology, coastal engineering, survey assessment etc.

- Bathymetric LiDAR: Bathymetric LiDARs are used in measuring the depth of water bodies. In a bathymetric LiDAR survey, the infrared light is reflected back to the aircraft from the land and water surface, while the additional green laser travels through the water column. Bathymetric information is crucial near coastlines, in harbors, and near shores and banks. Bathymetric information is also used to locate objects on the ocean floor.

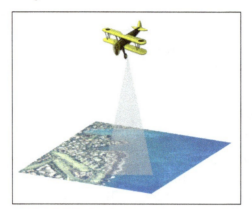

Terrestrial LiDAR

Terrestrial LiDAR can be installed either on a tripod or on a moving vehicle. It collects data points that help in the highly-accurate identification of data. This has its application in surveying and creating 3D Modeling. Terrestrial LiDAR can be either Mobile or Static.

- Mobile: It is mostly used to analyze infrastructure and observe roads. Mobile LiDAR systems mostly include sensors, camera and GPS.

- Static: Static LiDAR is more portable, handy to use. It collects cloud points from a fixed location and is used in mining and archaeology.

Based on other Classifications

- DIAL: DIAL is an acronym of Differential Absorption LiDAR sensing of ozone. It is mostly used to measure Ozone in the lower atmosphere.

- Raman LIDAR: It is used for profiling water vapor and aerosol.

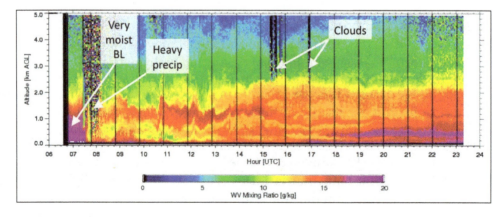

- Wind LiDAR: Wind LIDAR is used to measure wind speed and direction with high accuracy Wind data has been conventionally gathered with great difficulty due to multiple measurement points leading to inaccuracies. Using LiDAR one can measure wind speed, direction as well as turbulence.

- Spaceborne LiDAR: The potential of LiDAR extends beyond the earth as well. Premier space agencies including NASA are using LiDAR in detection and tracking.

- HSRL LiDAR: The NASA airborne High Spectral Resolution LiDAR (HSRL) is used to characterize clouds and small particles in the atmosphere, called aerosols. From an airborne platform, the HSRL scientist team analyzes aerosol size, composition, distribution and movement. The HSRL instrument is an innovative technology that is similar to radar; however, in the case of LiDAR, radio waves are replaced with laser light.

    The HSRL technique utilizes spectral distribution of the LiDAR return signal to discriminate aerosol and molecular signals and thereby measure aerosol extinction and backscatter independently.

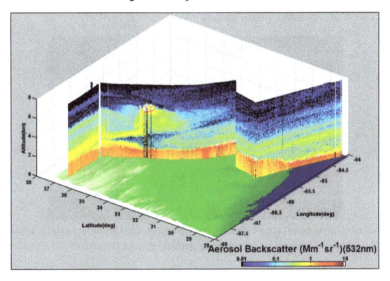

LiDAR Mapping

Light detection and ranging (lidar) is a remote sensing technology used to acquire elevation data about the Earth's surface. A lidar system consisted of three main components: the laser ranging system, Global Positioning System (GPS) and Inertial Measurement Unit (IMU). The laser ranging system transmits a laser pulse down to the Earth's surface, and the time delay between the transmission of the laser pulse and its return to the sensor is measured. This two-way travel time is then converted into a range or distance from the sensor to the target.

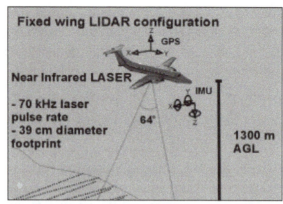

Typical lidar configuration used by AGRG for the topographic coastal surveys.

The range has a time stamp based on the precise GPS time and is related to the lidar sensor scan angle and the aircraft position. The aircrafts' 3-D (x, y, z) position is maintained by the precise survey GPS and the IMU is used to measure the roll, pitch and heading of the aircraft so that the ranges measured by the lidar sensor can be converted to points and georeferenced (x, y, z coordinates) and represented as a lidar point cloud). The point cloud is processed using software to detect points that represent the ground. This is known as lidar point classification.

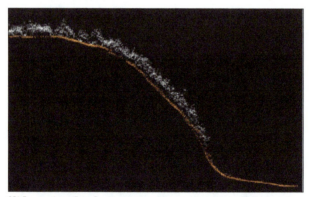

Example of lidar point cloud: Orange points represent ground points and white points represent non-ground points (trees).

How can we use Lidar Technologies for Flood Risk Mapping?

The minimum lidar classification involves detecting the lidar points that represent the ground. More advanced classifications involve determining if the points represent vegetation, building or power lines for example. Once the lidar is classified, the points are converted into a continuous surface model that can be colourized and have artificial sun shading applied to enhance the relief. If only the ground lidar points are used to produce the continuous surface we call that a "bare earth" Digital Elevation Model (DEM) or a Digital Terrain Model (DTM). If all of the valid lidar points are used to build the model it is knows as a Digital Surface Model (DSM). Since the DSM represents the tops of the building and trees, it can be subtracted from the DEM to produce a Normalized

Height Model (NHM), or if forest only a Canopy height Model (CHM). These models are stored in a raster format as grid cell or pixels where the value of each cell represents the elevation of the feature (eg. Ground or building etc). The surface models constructed from the lidar have grid cells of 1-2 m in size because of the dense point spacing resulting from the lidar survey. The vertical accuracy of the lidar data is also very high and is typically within 15-30 cm in open flat terrain, thus making it the preferred data to use for flood risk mapping.

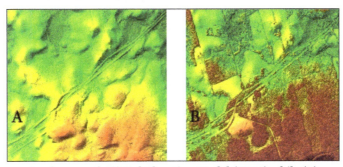

(A) Represents a bare earth Digital Elevation Model (DEM) while (B) represents a Digital Surface Model (DSM) including features such as buildings and trees.

The bare earth DEM's were used to generate the flood inundation layers for this project. By ensuring that the DEM's are hydraulically connected. Roads which have streams or rivers running under them most often contain a culvert or bridge to allow for the flow of water. The lidar is not able to measure below the earth and detect a culvert, therefore the road on the DEM does not represent the correct path for water to move. In these cases the DEM was modified and the values altered where culverts and bridges exist so that as the ocean water is raised, it can move to low lying areas correctly. Ensuring hydraulic connectivity allows flood inundation mapping to be as accurate as possible. The latest lidar systems allow elevation to be surveyed above and below the water line.

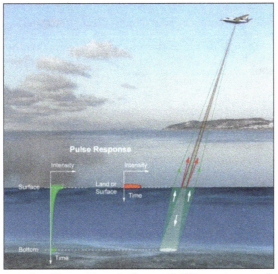

Example of how a topo-bathymetric lidar system works.

NSCC acquired a topographic-bathymetric lidar system. The Chiroptera II System manufactured by Leica Geosystems is equipped with 4 different sensors: a near-infrared (NIR) laser that can fire at 500kHz (500,000 shots per second) for mapping topography, a green laser (35kHz) for mapping bathymetry, and 60 megapixel camera that can acquire true colour (RGB) and near-infrared (NIR) images as well as a 5 megapixel quality assurance (QA) camera.

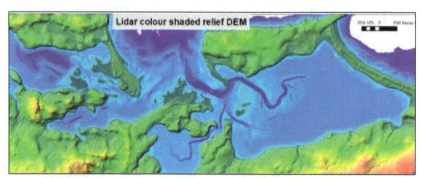

Example of a seamless DEM from a topo-bathymetric lidar sensor.

The green laser is able to penetrate through the water column to reach the bottom of the seabed, enabling the bathymetry to capture. This allows a continuous DEM to be constructed that seamlessly measures the elevation from land to under the sea or fresh water. The ability to survey the near-shore environments instead of just the land improves our ability to use hydrodynamic models to circulate the water during storm surges. The nearshore topography, water level and wind are the dominant factor controlling how waves behave near the shoreline. As more data is collected using this system, eventually wave runup can be modelled and incorporated into this flood risk mapping tool.

References

- Radar-basics-part-2-pulse-doppler-radar: eetimes.com, Retrieved 15, January 2020
- Radar-basics-types-and-applications: elprocus.com, Retrieved 04, August 2020
- What-is-lidar-technology-and-how-does-it-work: geospatialworld.net, Retrieved 29, March 2020
- Lidar-light-detection-and-ranging-working-application: elprocus.com, Retrieved 02, July 2020
- Do-you-know-types-of-lidar: geospatialworld.net, Retrieved 24, April 2020
- What-is-lidar-mapping, CoastalFlooding: cogs.nscc.ca, Retrieved 18, February 2020

Chapter 4
GIS and GPS

A computer system used for recording, storing, examining and displaying data related to positions on Earth's surface is known as Geographic Information System. Global Positioning System is a navigation system that provides geolocation and time information by using satellites, a receiver and algorithms. All the diverse principles of Geographic Information System have been carefully analyzed in this chapter.

Geographic Information System

A geographic information system or geographical information system (GIS) is a system designed to capture, store, manipulate, analyze, manage, and present all types of spatial or geographical data. The acronym GIS is sometimes used for geographic information science (GIScience) to refer to the academic discipline that studies geographic information systems and is a large domain within the broader academic discipline of geoinformatics. What goes beyond a GIS is a spatial data infrastructure, a concept that has no such restrictive boundaries.

In a general sense, the term describes any information system that integrates, stores, edits, analyzes, shares, and displays geographic information. GIS applications are tools that allow users to create interactive queries (user-created searches), analyze spatial information, edit data in maps, and present the results of all these operations. Geographic information science is the science underlying geographic concepts, applications, and systems.

GIS is a broad term that can refer to a number of different technologies, processes, and methods. It is attached to many operations and has many applications related to engineering, planning, management, transport/logistics, insurance, telecommunications, and business. For that reason, GIS and location intelligence applications can be the foundation for many location-enabled services that rely on analysis and visualization.

GIS can relate unrelated information by using location as the key index variable. Locations or extents in the Earth space–time may be recorded as dates/times of occurrence, and x, y, and z coordinates representing, longitude, latitude, and elevation, respectively. All Earth-based spatial–temporal location and extent references should, ideally, be relatable to one another and ultimately to a "real" physical location or extent. This key characteristic of GIS has begun to open new avenues of scientific inquiry.

History of Development

The first known use of the term "geographic information system" was by Roger Tomlinson in the year 1968 in his paper "A Geographic Information System for Regional Planning". Tomlinson is also acknowledged as the "father of GIS".

E. W. Gilbert's version (1958) of John Snow's 1855 map of the Soho cholera outbreak showing the clusters of cholera cases in the London epidemic of 1854.

Previously, one of the first applications of spatial analysis in epidemiology is the 1832 *"Rapport sur la marche et les effets du choléra dans Paris et le département de la Seine"*. The French geographer Charles Picquet represented the 48 districts of the city of Paris by halftone color gradient according to the number of deaths by cholera per 1,000 inhabitants. In 1854 John Snow determined the source of a cholera outbreak in London by marking points on a map depicting where the cholera victims lived, and connecting the cluster that he found with a nearby water source. This was one of the earliest successful uses of a geographic methodology in epidemiology. While the basic elements of topography and theme existed previously in cartography, the John Snow map was unique, using cartographic methods not only to depict but also to analyze clusters of geographically dependent phenomena.

The early 20th century saw the development of photozincography, which allowed maps to be split into layers, for example one layer for vegetation and another for water. This was particularly used for printing contours – drawing these was a labour-intensive task but having them on a separate layer meant they could be worked on without the other layers to confuse the draughtsman. This work was originally drawn on glass plates but later plastic film was introduced, with the advantages of being lighter, using less storage space and being less brittle, among others. When all the layers were finished, they were combined into one image using a large process camera. Once color printing came in, the layers idea was also used for creating separate printing plates for each color. While the use of layers much later became one of the main typical features of a contemporary GIS, the photographic process just described is not considered to be a GIS in itself – as the maps were just images with no database to link them to.

Computer hardware development spurred by nuclear weapon research led to general-purpose computer "mapping" applications by the early 1960s.

The year 1960 saw the development of the world's first true operational GIS in Ottawa, Ontario, Canada by the federal Department of Forestry and Rural Development. Developed by Dr. Roger Tomlinson, it was called the Canada Geographic Information System (CGIS) and was used to store, analyze, and manipulate data collected for the Canada Land Inventory – an effort to determine the land capability for rural Canada by mapping information about soils, agriculture, recreation, wildlife, waterfowl, forestry and land use at a scale of 1:50,000. A rating classification factor was also added to permit analysis.

CGIS was an improvement over "computer mapping" applications as it provided capabilities for overlay, measurement, and digitizing/scanning. It supported a national coordinate system that spanned the continent, coded lines as arcs having a true embedded topology and it stored the attribute and locational information in separate files. As a result of this, Tomlinson has become known as the "father of GIS", particularly for his use of overlays in promoting the spatial analysis of convergent geographic data.

CGIS lasted into the 1990s and built a large digital land resource database in Canada. It was developed as a mainframe-based system in support of federal and provincial resource planning and management. Its strength was continent-wide analysis of complex datasets. The CGIS was never available commercially.

In 1964 Howard T. Fisher formed the Laboratory for Computer Graphics and Spatial Analysis at the Harvard Graduate School of Design (LCGSA 1965–1991), where a number of important theoretical concepts in spatial data handling were developed, and which by the 1970s had distributed seminal software code and systems, such as SYMAP, GRID, and ODYSSEY – that served as sources for subsequent commercial development—to universities, research centers and corporations worldwide.

By the late 1970s two public domain GIS systems (MOSS and GRASS GIS) were in development, and by the early 1980s, M&S Computing (later Intergraph) along with Bentley Systems Incorporated for the CAD platform, Environmental Systems Research Institute (ESRI), CARIS (Computer Aided Resource Information System), MapInfo Corporation and ERDAS (Earth Resource Data Analysis System) emerged as commercial vendors of GIS software, successfully incorporating many of the CGIS features, combining the first generation approach to separation of spatial and attribute information with a second generation approach to organizing attribute data into database structures.

In 1986, Mapping Display and Analysis System (MIDAS), the first desktop GIS product emerged for the DOS operating system. This was renamed in 1990 to MapInfo for Windows when it was ported to the Microsoft Windows platform. This began the process of moving GIS from the research department into the business environment.

By the end of the 20th century, the rapid growth in various systems had been consolidated and standardized on relatively few platforms and users were beginning to explore viewing GIS data over the Internet, requiring data format and transfer standards. More recently, a growing number of free, open-source GIS packages run on a range of operating systems and can be customized to perform specific tasks. Increasingly geospatial data and mapping applications are being made available via the world wide web.

Several articles on the history of GIS have been published.

GIS Techniques and Technology

Modern GIS technologies use digital information, for which various digitized data creation methods are used. The most common method of data creation is digitization, where a hard copy map or survey plan is transferred into a digital medium through the use of a CAD program, and geo-referencing capabilities. With the wide availability of ortho-rectified imagery (from satellites, aircraft, Helikites and UAVs), heads-up digitizing is becoming the main avenue through which geographic data is extracted. Heads-up digitizing involves the tracing of geographic data directly on top of the aerial imagery instead of by the traditional method of tracing the geographic form on a separate digitizing tablet (heads-down digitizing).

Relating Information from Different Sources

GIS uses spatio-temporal (space-time) location as the key index variable for all other information. Just as a relational database containing text or numbers can relate many different tables using common key index variables, GIS can relate otherwise unrelated information by using location as the key index variable. The key is the location and/or extent in space-time.

Any variable that can be located spatially, and increasingly also temporally, can be referenced using a GIS. Locations or extents in Earth space–time may be recorded as dates/times of occurrence, and x, y, and z coordinates representing, longitude, latitude, and elevation, respectively. These GIS coordinates may represent other quantified systems of temporo-spatial reference (for example, film frame number, stream gage station, highway mile-marker, surveyor benchmark, building address, street intersection, entrance gate, water depth sounding, POS or CAD drawing origin/units). Units applied to recorded temporal-spatial data can vary widely, but all Earth-based spatial–temporal location and extent references should, ideally, be relatable to one another and ultimately to a "real" physical location or extent in space–time.

Related by accurate spatial information, an incredible variety of real-world and projected past or future data can be analyzed, interpreted and represented. This key characteristic of GIS has begun to open new avenues of scientific inquiry into behaviors and patterns of real-world information that previously had not been systematically correlated.

GIS Uncertainties

GIS accuracy depends upon source data, and how it is encoded to be data referenced. Land surveyors have been able to provide a high level of positional accuracy utilizing the GPS-derived positions. High-resolution digital terrain and aerial imagery, powerful computers and Web technology are changing the quality, utility, and expectations of GIS to serve society on a grand scale, but nevertheless there are other source data that affect overall GIS accuracy like paper maps, though these may be of limited use in achieving the desired accuracy.

In developing a digital topographic database for a GIS, topographical maps are the main source, and aerial photography and satellite imagery are extra sources for collecting data and identifying attributes which can be mapped in layers over a location facsimile of scale. The scale of a map and geographical rendering area representation type are very important aspects since the information content depends mainly on the scale set and resulting locatability of the map's representations. In order to digitize a map, the map has to be checked within theoretical dimensions, then scanned into a raster format, and resulting raster data has to be given a theoretical dimension by a rubber sheeting/warping technology process.

A quantitative analysis of maps brings accuracy issues into focus. The electronic and other equipment used to make measurements for GIS is far more precise than the machines of conventional map analysis. All geographical data are inherently inaccurate, and these inaccuracies will propagate through GIS operations in ways that are difficult to predict.

Data Representation

GIS data represents real objects (such as roads, land use, elevation, trees, waterways, etc.) with digital data determining the mix. Real objects can be divided into two abstractions: discrete objects (e.g., a house) and continuous fields (such as rainfall amount, or elevations). Traditionally, there are two broad methods used to store data in a GIS for both kinds of abstractions mapping references: raster images and vector. Points, lines, and polygons are the stuff of mapped location attribute references. A new hybrid method of storing data is that of identifying point clouds, which combine three-dimensional points with RGB information at each point, returning a "3D color image". GIS thematic maps then are becoming more and more realistically visually descriptive of what they set out to show or determine.

For a list of popular GIS file formats, such as shapefiles.

Data Capture

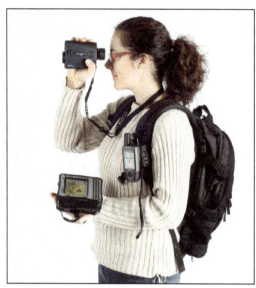

Example of hardware for mapping (GPS and laser rangefinder) and data collection (rugged computer). The current trend for geographical information system (GIS) is that accurate mapping and data analysis are completed while in the field. Depicted hardware (field-map technology) is used mainly for forest inventories, monitoring and mapping.

Data capture—entering information into the system—consumes much of the time of GIS practitioners. There are a variety of methods used to enter data into a GIS where it is stored in a digital format.

Existing data printed on paper or PET film maps can be digitized or scanned to produce digital data. A digitizer produces vector data as an operator traces points, lines, and polygon boundaries from a map. Scanning a map results in raster data that could be further processed to produce vector data.

Survey data can be directly entered into a GIS from digital data collection systems on survey instruments using a technique called coordinate geometry (COGO). Positions from a global navigation satellite system (GNSS) like Global Positioning System can also be collected and then imported into a GIS. A current trend in data collection gives users the ability to utilize field computers with the ability to edit live data using wireless connections or disconnected editing sessions. This has been enhanced by the availability of low-cost mapping-grade GPS units with decimeter accuracy in real time. This eliminates the need to post process, import, and update the data in the office after fieldwork has been collected. This includes the ability to incorporate positions collected using a laser rangefinder. New technologies also allow users to create maps as well as analysis directly in the field, making projects more efficient and mapping more accurate.

Remotely sensed data also plays an important role in data collection and consist of sensors attached to a platform. Sensors include cameras, digital scanners and lidar, while platforms usually consist of aircraft and satellites. In England in the mid 1990s, hybrid kite/balloons called Helikites first pioneered the use of compact airborne digital cameras as airborne Geo-Information Systems. Aircraft measurement software, accurate to 0.4 mm was used to link the photographs and measure the ground. Helikites are inexpensive and gather more accurate data than aircraft. Helikites can be used over roads, railways and towns where UAVs are banned.

Recently with the development of miniature UAVs, aerial data collection is becoming possible with them. For example, the Aeryon Scout was used to map a 50-acre area with a Ground sample distance of 1 inch (2.54 cm) in only 12 minutes.

The majority of digital data currently comes from photo interpretation of aerial photographs. Soft-copy workstations are used to digitize features directly from stereo pairs of digital photographs. These systems allow data to be captured in two and three dimensions, with elevations measured directly from a stereo pair using principles of photogrammetry. Analog aerial photos must be scanned before being entered into a soft-copy system, for high-quality digital cameras this step is skipped.

Satellite remote sensing provides another important source of spatial data. Here satellites use different sensor packages to passively measure the reflectance from parts of the electromagnetic spectrum or radio waves that were sent out from an active sensor such as radar. Remote sensing collects raster data that can be further processed using different bands to identify objects and classes of interest, such as land cover.

When data is captured, the user should consider if the data should be captured with either a relative accuracy or absolute accuracy, since this could not only influence how information will be interpreted but also the cost of data capture.

After entering data into a GIS, the data usually requires editing, to remove errors, or further processing. For vector data it must be made "topologically correct" before it can be used for some advanced analysis. For example, in a road network, lines must connect with nodes at an intersection. Errors such as undershoots and overshoots must also be removed. For scanned maps, blemishes on the source map may need to be removed from the resulting raster. For example, a fleck of dirt might connect two lines that should not be connected.

Raster-to-vector Translation

Data restructuring can be performed by a GIS to convert data into different formats. For example, a GIS may be used to convert a satellite image map to a vector structure by generating lines around all cells with the same classification, while determining the cell spatial relationships, such as adjacency or inclusion.

More advanced data processing can occur with image processing, a technique developed in the late 1960s by NASA and the private sector to provide contrast enhancement, false color rendering and a variety of other techniques including use of two dimensional Fourier transforms. Since digital data is collected and stored in various ways, the two data sources may not be entirely compatible. So a GIS must be able to convert geographic data from one structure to another. In so doing, the implicit assumptions behind different ontologies and classifications require analysis. Object ontologies have gained increasing prominence as a consequence of object-oriented programming and sustained work by Barry Smith and co-workers.

Projections, Coordinate Systems, and Registration

The earth can be represented by various models, each of which may provide a different set of coordinates (e.g., latitude, longitude, elevation) for any given point on the Earth's surface. The simplest model is to assume the earth is a perfect sphere. As more measurements of the earth have accumulated, the models of the earth have become more sophisticated and more accurate. In fact, there are models called datums that apply to different areas of the earth to provide increased accuracy, like NAD83 for U.S. measurements, and the World Geodetic System for worldwide measurements.

Spatial Analysis with Geographical Information System (GIS)

GIS spatial analysis is a rapidly changing field, and GIS packages are increasingly including analytical tools as standard built-in facilities, as optional toolsets, as add-ins or 'analysts'. In many instances these are provided by the original software suppliers (commercial vendors or collaborative non commercial development teams), while in other cases facilities have been developed and are provided by third parties. Furthermore, many products offer software development kits (SDKs), programming languages and language support, scripting facilities and/or special interfaces for developing one's own analytical tools or variants. The website "Geospatial Analysis" and associated book/ebook attempt to provide a reasonably comprehensive guide to the subject. The increased availability has created a new dimension to business intelligence termed "spatial intelligence" which, when openly delivered via intranet, democratizes access to geographic and social network data. Geospatial intelligence, based on GIS spatial analysis, has also become a key element for security. GIS as a whole can be described as conversion to a vectorial representation or to any other digitisation process.

Slope and Aspect

Slope can be defined as the steepness or gradient of a unit of terrain, usually measured as an angle in degrees or as a percentage. Aspect can be defined as the direction in which a unit of terrain faces. Aspect is usually expressed in degrees from north. Slope, aspect, and surface curvature in terrain analysis are all derived from neighborhood operations using elevation values of a cell's adjacent neighbours. Slope is a function of resolution, and the spatial resolution used to calculate slope and aspect should always be specified. Authors such as Skidmore, Jones and Zhou and Liu have compared techniques for calculating slope and aspect.

The following method can be used to derive slope and aspect:

The elevation at a point or unit of terrain will have perpendicular tangents (slope) passing through the point, in an east-west and north-south direction. These two tangents give two components, $\partial z/\partial x$ and $\partial z/\partial y$, which then be used to determine the overall direction of slope, and the aspect of the slope. The gradient is defined as a vector quantity with components equal to the partial derivatives of the surface in the x and y directions.

The calculation of the overall 3x3 grid slope S and aspect A for methods that determine east-west and north-south component use the following formulas respectively:

$$\tan S = \sqrt{\left(\frac{\partial z}{\partial x}\right)^2 + \left(\frac{\partial z}{\partial y}\right)^2}$$

$$\tan A = \left(\frac{\left(\frac{-\partial z}{\partial y}\right)}{\left(\frac{\partial z}{\partial x}\right)}\right)$$

Zhou and Liu describe another formula for calculating aspect, as follows:

$$A = 270° + \arctan\left(\frac{\left(\frac{\partial z}{\partial x}\right)}{\left(\frac{\partial z}{\partial y}\right)}\right) - 90°\left(\frac{\left(\frac{\partial z}{\partial y}\right)}{\left|\frac{\partial z}{\partial y}\right|}\right)$$

Data Analysis

It is difficult to relate wetlands maps to rainfall amounts recorded at different points such as airports, television stations, and schools. A GIS, however, can be used to depict two- and three-dimensional characteristics of the Earth's surface, subsurface, and atmosphere from information points. For example, a GIS can quickly generate a map with isopleth or contour lines that indicate differing amounts of rainfall. Such a map can be thought of as a rainfall contour map. Many sophisticated methods can estimate the characteristics of surfaces from a limited number of point measurements. A two-dimensional contour map created from the surface modeling of rainfall point measurements may be overlaid and analyzed with any other map in a GIS covering the same area. This GIS derived map can then provide additional information - such as the viability of water power potential as a renewable energy source. Similarly, GIS can be used to compare other renewable energy resources to find the best geographic potential for a region.

Additionally, from a series of three-dimensional points, or digital elevation model, isopleth lines representing elevation contours can be generated, along with slope analysis, shaded relief, and other elevation products. Watersheds can be easily defined for any given reach, by computing all of the areas contiguous and uphill from any given point of interest. Similarly, an expected thalweg of where surface water would want to travel in intermittent and permanent streams can be computed from elevation data in the GIS.

Topological Modeling

A GIS can recognize and analyze the spatial relationships that exist within digitally stored spatial data. These topological relationships allow complex spatial modelling and analysis to be performed. Topological relationships between geometric entities traditionally include adjacency (what adjoins what), containment (what encloses what), and proximity (how close something is to something else).

Geometric Networks

Geometric networks are linear networks of objects that can be used to represent interconnected features, and to perform special spatial analysis on them. A geometric network is composed of edges, which are connected at junction points, similar to graphs in mathematics and computer science. Just like graphs, networks can have weight and flow assigned to its edges, which can be used to represent various interconnected features more accurately. Geometric networks are often used to model road networks and public utility networks, such as electric, gas, and water networks. Network modeling is also commonly employed in transportation planning, hydrology modeling, and infrastructure modeling.

Hydrological Modeling

GIS hydrological models can provide a spatial element that other hydrological models lack, with the analysis of variables such as slope, aspect and watershed or catchment area. Terrain analysis is fundamental to hydrology, since water always flows down a slope. As basic terrain analysis of a digital elevation model (DEM) involves calculation of slope and aspect, DEMs are very useful for hydrological analysis. Slope and aspect can then be used to determine direction of surface runoff, and hence flow accumulation for the formation of streams, rivers and lakes. Areas of divergent flow can also give a clear indication of the boundaries of a catchment. Once a flow direction and accumulation matrix has been created, queries can be performed that show contributing or dispersal areas at a certain point. More detail can be added to the model, such as terrain roughness, vegetation types and soil types, which can influence infiltration and evapotranspiration rates, and hence influencing surface flow. One of the main uses of hydrological modeling is in environmental contamination research.

Cartographic Modeling

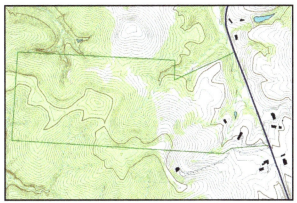

An example of use of layers in a GIS application. In this example, the forest cover layer (light green) is at the bottom, with the topographic layer over it. Next up is the stream layer, then the boundary layer, then the road layer. The order is very important in order to properly display the final result. Note that the pond layer was located just below the stream layer, so that a stream line can be seen overlying one of the ponds.

The term "cartographic modeling" was probably coined by Dana Tomlin in his PhD dissertation and later in his book which has the term in the title. Cartographic modeling refers to a process where several thematic layers of the same area are produced, processed, and analyzed. Tomlin used raster layers, but the overlay method can be used more generally. Operations on map layers can be combined into algorithms, and eventually into simulation or optimization models.

Map Overlay

The combination of several spatial datasets (points, lines, or polygons) creates a new output vector dataset, visually similar to stacking several maps of the same region. These overlays are similar to mathematical Venn diagram overlays. A union overlay combines the geographic features and attribute tables of both inputs into a single new output. An intersect overlay defines the area where both inputs overlap and retains a set of attribute fields for each. A symmetric difference overlay defines an output area that includes the total area of both inputs except for the overlapping area.

Data extraction is a GIS process similar to vector overlay, though it can be used in either vector or raster data analysis. Rather than combining the properties and features of both datasets, data extraction involves using a "clip" or "mask" to extract the features of one data set that fall within the spatial extent of another dataset.

In raster data analysis, the overlay of datasets is accomplished through a process known as "local operation on multiple rasters" or "map algebra," through a function that combines the values of each raster's matrix. This function may weigh some inputs more than others through use of an "index model" that reflects the influence of various factors upon a geographic phenomenon.

Geostatistics

Geostatistics is a branch of statistics that deals with field data, spatial data with a continuous index. It provides methods to model spatial correlation, and predict values at arbitrary locations (interpolation).

When phenomena are measured, the observation methods dictate the accuracy of any subsequent analysis. Due to the nature of the data (e.g. traffic patterns in an urban environment; weather patterns over the Pacific Ocean), a constant or dynamic degree of precision is always lost in the measurement. This loss of precision is determined from the scale and distribution of the data collection.

To determine the statistical relevance of the analysis, an average is determined so that points (gradients) outside of any immediate measurement can be included to determine their predicted behavior. This is due to the limitations of the applied statistic and data collection methods, and interpolation is required to predict the behavior of particles, points, and locations that are not directly measurable.

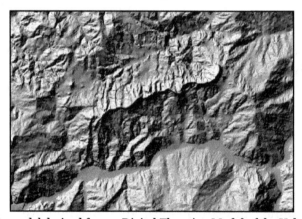

Hillshade model derived from a Digital Elevation Model of the Valestra area in the northern Apennines (Italy).

Interpolation is the process by which a surface is created, usually a raster dataset, through the input of data collected at a number of sample points. There are several forms of interpolation, each which treats the data differently, depending on the properties of the data set. In comparing interpolation methods, the first consideration should be whether or not the source data will change (exact or approximate). Next is whether the method is subjective, a human interpretation, or objective. Then there is the nature of transitions between points: are they abrupt or gradual. Finally, there is whether a method is global (it uses the entire data set to form the model), or local where an algorithm is repeated for a small section of terrain.

Interpolation is a justified measurement because of a spatial autocorrelation principle that recognizes that data collected at any position will have a great similarity to, or influence of those locations within its immediate vicinity.

Digital elevation models, triangulated irregular networks, edge-finding algorithms, Thiessen polygons, Fourier analysis, (weighted) moving averages, inverse distance weighting, kriging, spline, and trend surface analysis are all mathematical methods to produce interpolative data.

Address Geocoding

Geocoding is interpolating spatial locations (X,Y coordinates) from street addresses or any other spatially referenced data such as ZIP Codes, parcel lots and address locations. A reference theme is required to geocode individual addresses, such as a road centerline file with address ranges. The individual address locations have historically been interpolated, or estimated, by examining address ranges along a road segment. These are usually provided in the form of a table or database. The software will then place a dot approximately where that address belongs along the segment of centerline. For example, an address point of 500 will be at the midpoint of a line segment that starts with address 1 and ends with address 1,000. Geocoding can also be applied against actual parcel data, typically from municipal tax maps. In this case, the result of the geocoding will be an actually positioned space as opposed to an interpolated point. This approach is being increasingly used to provide more precise location information.

Reverse Geocoding

Reverse geocoding is the process of returning an estimated street address number as it relates to a given coordinate. For example, a user can click on a road centerline theme (thus providing a coordinate) and have information returned that reflects the estimated house number. This house number is interpolated from a range assigned to that road segment. If the user clicks at the midpoint of a segment that starts with address 1 and ends with 100, the returned value will be somewhere near 50. Note that reverse geocoding does not return actual addresses, only estimates of what should be there based on the predetermined range.

Multi-criteria Decision Analysis

Coupled with GIS, multi-criteria decision analysis methods support decision-makers in analysing a set of alternative spatial solutions, such as the most likely ecological habitat for restoration, against multiple criteria, such as vegetation cover or roads. MCDA uses decision rules to aggregate the criteria, which allows the alternative solutions to be ranked or prioritised. GIS MCDA may reduce costs and time involved in identifying potential restoration sites.

Data Output and Cartography

Cartography is the design and production of maps, or visual representations of spatial data. The vast majority of modern cartography is done with the help of computers, usually using GIS but production of quality cartography is also achieved by importing layers into a design program to refine it. Most GIS software gives the user substantial control over the appearance of the data.

Cartographic work serves two major functions:

First, it produces graphics on the screen or on paper that convey the results of analysis to the people who make decisions about resources. Wall maps and other graphics can be generated, allowing the viewer to visualize and thereby understand the results of analyses or simulations of potential events. Web Map Servers facilitate distribution of generated maps through web browsers using various implementations of web-based application programming interfaces (AJAX, Java, Flash, etc.).

Second, other database information can be generated for further analysis or use. An example would be a list of all addresses within one mile (1.6 km) of a toxic spill.

Graphic Display Techniques

Traditional maps are abstractions of the real world, a sampling of important elements portrayed on a sheet of paper with symbols to represent physical objects. People who use maps must interpret these symbols. Topographic maps show the shape of land surface with contour lines or with shaded relief.

Today, graphic display techniques such as shading based on altitude in a GIS can make relationships among map elements visible, heightening one's ability to extract and analyze information. For example, two types of data were combined in a GIS to produce a perspective view of a portion of San Mateo County, California.

- The digital elevation model, consisting of surface elevations recorded on a 30-meter horizontal grid, shows high elevations as white and low elevation as black.

- The accompanying Landsat Thematic Mapper image shows a false-color infrared image looking down at the same area in 30-meter pixels, or picture elements, for the same coordinate points, pixel by pixel, as the elevation information.

A GIS was used to register and combine the two images to render the three-dimensional perspective view looking down the San Andreas Fault, using the Thematic Mapper image pixels, but shaded using the elevation of the landforms. The GIS display depends on the viewing point of the observer and time of day of the display, to properly render the shadows created by the sun's rays at that latitude, longitude, and time of day.

An archeochrome is a new way of displaying spatial data. It is a thematic on a 3D map that is applied to a specific building or a part of a building. It is suited to the visual display of heat-loss data.

Spatial ETL

Spatial ETL tools provide the data processing functionality of traditional Extract, Transform, Load (ETL) software, but with a primary focus on the ability to manage spatial data. They provide GIS users with the ability to translate data between different standards and proprietary formats, whilst geometrically transforming the data en route. These tools can come in the form of add-ins to existing wider-purpose software such as Microsoft Excel.

GIS Data Mining

GIS or spatial data mining is the application of data mining methods to spatial data. Data mining, which is the partially automated search for hidden patterns in large databases, offers great potential benefits for applied GIS-based decision making. Typical applications include environmental monitoring. A characteristic of such applications is that spatial correlation between data measurements require the use of specialized algorithms for more efficient data analysis.

Applications

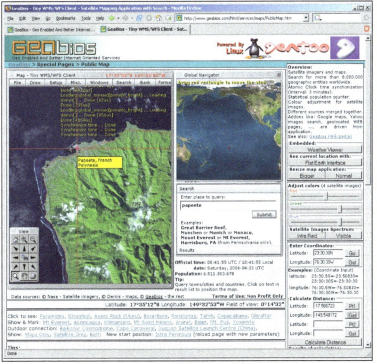

GeaBios – tiny WMS/WFS client (Flash/DHTML).

The implementation of a GIS is often driven by jurisdictional (such as a city), purpose, or application requirements. Generally, a GIS implementation may be custom-designed for an organization. Hence, a GIS deployment developed for an application, jurisdiction, enterprise, or purpose may not be necessarily interoperable or compatible with a GIS that has been developed for some other application, jurisdiction, enterprise, or purpose.

GIS provides, for every kind of location-based organization, a platform to update geographical data without wasting time to visit the field and update a database manually. GIS when integrated with other powerful enterprise solutions like SAP and the Wolfram Language helps creating powerful decision support system at enterprise level.

Many disciplines can benefit from GIS technology. An active GIS market has resulted in lower costs and continual improvements in the hardware and software components of GIS, and usage in the fields of science, government, business, and industry, with applications including real estate, public health, crime mapping, national defense, sustainable development, natural resources, climatology, landscape architecture, archaeology, regional and community planning, transportation and logistics. GIS is also diverging into location-based services, which allows GPS-enabled mobile devices to display their location in relation to fixed objects (nearest restaurant, gas station, fire hydrant) or mobile objects (friends, children, police car), or to relay their position back to a central server for display or other processing.

Open Geospatial Consortium Standards

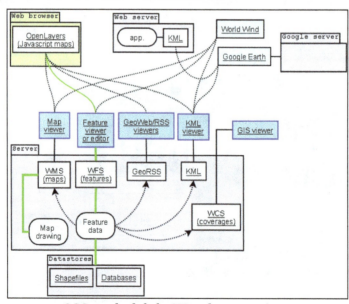

OGC standards help GIS tools communicate.

The Open Geospatial Consortium (OGC) is an international industry consortium of 384 companies, government agencies, universities, and individuals participating in a consensus process to develop publicly available geoprocessing specifications. Open interfaces and protocols defined by OpenGIS Specifications support interoperable solutions that "geo-enable" the Web, wireless and location-based services, and mainstream IT, and empower technology developers to make complex spatial information and services accessible and useful with all kinds of applications. Open Geospatial Consortium protocols include Web Map Service, and Web Feature Service.

GIS products are broken down by the OGC into two categories, based on how completely and accurately the software follows the OGC specifications.

Compliant Products are software products that comply to OGC's OpenGIS Specifications. When a product has been tested and certified as compliant through the OGC Testing Program, the product is automatically registered as "compliant" on this site.

Implementing Products are software products that implement OpenGIS Specifications but have not yet passed a compliance test. Compliance tests are not available for all specifications. Developers can register their products as implementing draft or approved specifications, though OGC reserves the right to review and verify each entry.

Web Mapping

In recent years there has been an explosion of mapping applications on the web such as Google Maps and Bing Maps. These websites give the public access to huge amounts of geographic data.

Some of them, like Google Maps and OpenLayers, expose an API that enable users to create custom applications. These toolkits commonly offer street maps, aerial/satellite imagery, geocoding, searches, and routing functionality. Web mapping has also uncovered the potential of crowdsourcing geodata in projects like OpenStreetMap, which is a collaborative project to create a free editable map of the world.

Adding the Dimension of Time

The condition of the Earth's surface, atmosphere, and subsurface can be examined by feeding satellite data into a GIS. GIS technology gives researchers the ability to examine the variations in Earth processes over days, months, and years. As an example, the changes in vegetation vigor through a growing season can be animated to determine when drought was most extensive in a particular region. The resulting graphic represents a rough measure of plant health. Working with two variables over time would then allow researchers to detect regional differences in the lag between a decline in rainfall and its effect on vegetation.

GIS technology and the availability of digital data on regional and global scales enable such analyses. The satellite sensor output used to generate a vegetation graphic is produced for example by the Advanced Very High Resolution Radiometer (AVHRR). This sensor system detects the amounts of energy reflected from the Earth's surface across various bands of the spectrum for surface areas of about 1 square kilometer. The satellite sensor produces images of a particular location on the Earth twice a day. AVHRR and more recently the Moderate-Resolution Imaging Spectroradiometer (MODIS) are only two of many sensor systems used for Earth surface analysis. More sensors will follow, generating ever greater amounts of data.

In addition to the integration of time in environmental studies, GIS is also being explored for its ability to track and model the progress of humans throughout their daily routines. A concrete example of progress in this area is the recent release of time-specific population data by the U.S. Census. In this data set, the populations of cities are shown for daytime and evening hours highlighting the pattern of concentration and dispersion generated by North American commuting patterns. The manipulation and generation of data required to produce this data would not have been possible without GIS.

Using models to project the data held by a GIS forward in time have enabled planners to test policy decisions using spatial decision support systems.

Semantics

Tools and technologies emerging from the W3C's Data Activity are proving useful for data integration problems in information systems. Correspondingly, such technologies have been proposed as a means to facilitate interoperability and data reuse among GIS applications. and also to enable new analysis mechanisms.

Ontologies are a key component of this semantic approach as they allow a formal, machine-readable specification of the concepts and relationships in a given domain. This in turn allows a GIS to focus on the intended meaning of data rather than its syntax or structure. For example, reasoning that a land cover type classified as *deciduous needleleaf trees* in one dataset is a specialization or subset of land cover type *forest* in another more roughly classified dataset can help a GIS automatically merge the two datasets under the more general land cover classification. Tentative ontologies have been developed in areas related to GIS applications, for example the hydrology ontology developed by the Ordnance Survey in the United Kingdom and the SWEET ontologies developed by NASA's Jet Propulsion Laboratory. Also, simpler ontologies and semantic metadata standards are being proposed by the W3C Geo Incubator Group to represent geospatial data on the web. GeoSPARQL is a standard developed by the Ordnance Survey, United States Geological Survey, Natural Resources Canada, Australia's Commonwealth Scientific and Industrial Research Organisation and others to support ontology creation and reasoning using well-understood OGC literals (GML, WKT), topological relationships (Simple Features, RCC8, DE-9IM), RDF and the SPARQL database query protocols.

Recent research results in this area can be seen in the International Conference on Geospatial Semantics and the Terra Cognita – Directions to the Geospatial Semantic Web workshop at the International Semantic Web Conference.

Implications of GIS in Society

With the popularization of GIS in decision making, scholars have begun to scrutinize the social and political implications of GIS. GIS can also be misused to distort reality for individual and political gain. It has been argued that the production, distribution, utilization, and representation of geographic information are largely related with the social context and has the potential to increase citizen trust in government. Other related topics include discussion on copyright, privacy, and censorship. A more optimistic social approach to GIS adoption is to use it as a tool for public participation.

Global Positioning System

Artist's conception of GPS Block II-F satellite in Earth orbit.

Civilian GPS receivers ("GPS navigation device") in a marine application.

The Global Positioning System (GPS), also known as Navstar, is a global navigation satellite system (GNSS) that provides location and time information in all weather conditions, anywhere on or near the Earth where there is an unobstructed line of sight to four or more GPS satellites. The GPS system operates independently of any telephonic or internet reception, though these technologies can enhance the usefulness of the GPS positioning information. The GPS system provides critical positioning capabilities to military, civil, and commercial users around the world. The United States government created the system, maintains it, and makes it freely accessible to anyone with a GPS receiver.

The United States began the GPS project in 1973 to overcome the limitations of previous navigation systems, integrating ideas from several predecessors, including a number of classified engineering design studies from the 1960s. The U.S. Department of Defense (DoD) developed the system, which originally used 24 satellites. It became fully operational in 1995. Roger L. Easton, Ivan A. Getting and Bradford Parkinson of the Applied Physics Laboratory are credited with inventing it.

Advances in technology and new demands on the existing system have now led to efforts to modernize the GPS and implement the next generation of GPS Block IIIA satellites and Next Generation Operational Control System (OCX). Announcements from Vice President Al Gore and the White House in 1998 initiated these changes. In 2000, the U.S. Congress authorized the modernization effort, GPS III.

In addition to GPS, other systems are in use or under development. The Russian Global Navigation Satellite System (GLONASS) was developed contemporaneously with GPS, but suffered from incomplete coverage of the globe until the mid-2000s. There are also the planned European Union Galileo positioning system, China's BeiDou Navigation Satellite System, the Japanese Quasi-Zenith Satellite System, and India's Indian Regional Navigation Satellite System.

History

The design of GPS is based partly on similar ground-based radio-navigation systems, such as LORAN and the Decca Navigator, developed in the early 1940s and used by the British Royal Navy during World War II.

In 1956, the German-American physicist Friedwardt Winterberg proposed a test of general relativity — detecting time slowing in a strong gravitational field using accurate atomic clocks placed in orbit inside artificial satellites.

Special and general relativity predict that the clocks on the GPS satellites would be seen by the Earth's observers to run 38 microseconds faster per day than the clocks on the Earth. The GPS calculated positions would quickly drift into error, accumulating to 10 kilometers per day. The relativistic time effect of the GPS clocks running faster than the clocks on earth was corrected for in the design of GPS.

Predecessors

The Soviet Union launched the first man-made satellite, Sputnik 1, in 1957. Two American physicists, William Guier and George Weiffenbach, at Johns Hopkins's Applied

Physics Laboratory (APL), decided to monitor Sputnik's radio transmissions. Within hours they realized that, because of the Doppler effect, they could pinpoint where the satellite was along its orbit. The Director of the APL gave them access to their UNIVAC to do the heavy calculations required.

The next spring, Frank McClure, the deputy director of the APL, asked Guier and Weiffenbach to investigate the inverse problem — pinpointing the user's location, given that of the satellite. (At the time, the Navy was developing the submarine-launched Polaris missile, which required them to know the submarine's location.) This led them and APL to develop the TRANSIT system. In 1959, ARPA (renamed DARPA in 1972) also played a role in TRANSIT.

Official logo for NAVSTAR GPS.

Emblem of the 50th Space Wing.

The first satellite navigation system, TRANSIT, used by the United States Navy, was first successfully tested in 1960. It used a constellation of five satellites and could provide a navigational fix approximately once per hour.

In 1967, the U.S. Navy developed the Timation satellite that proved the ability to place accurate clocks in space, a technology required by GPS.

In the 1970s, the ground-based OMEGA navigation system, based on phase comparison of signal transmission from pairs of stations, became the first worldwide radio navigation system. Limitations of these systems drove the need for a more universal navigation solution with greater accuracy.

While there were wide needs for accurate navigation in military and civilian sectors, almost none of those was seen as justification for the billions of dollars it would cost in research, development, deployment, and operation for a constellation of navigation satellites. During the Cold War arms race, the nuclear threat to the existence of the United States was the one need that did justify this cost in the view of the United States Congress. This deterrent effect is why GPS was funded. It is also the reason for the ultra secrecy at that time. The nuclear triad consisted of the United States Navy's submarine-launched ballistic missiles (SLBMs) along with United States Air Force (USAF) strategic bombers and intercontinental ballistic missiles (ICBMs). Considered vital to the nuclear deterrence posture, accurate determination of the SLBM launch position was a force multiplier.

Precise navigation would enable United States ballistic missile submarines to get an accurate fix of their positions before they launched their SLBMs. The USAF, with two thirds of the nuclear triad, also had requirements for a more accurate and reliable navigation system. The Navy and Air Force were developing their own technologies in parallel to solve what was essentially the same problem.

To increase the survivability of ICBMs, there was a proposal to use mobile launch platforms (such as Russian SS-24 and SS-25) and so the need to fix the launch position had similarity to the SLBM situation.

In 1960, the Air Force proposed a radio-navigation system called MOSAIC (MObile System for Accurate ICBM Control) that was essentially a 3-D LORAN. A follow-on study, Project 57, was worked in 1963 and it was "in this study that the GPS concept was born." That same year, the concept was pursued as Project 621B, which had "many of the attributes that you now see in GPS" and promised increased accuracy for Air Force bombers as well as ICBMs.

Updates from the Navy TRANSIT system were too slow for the high speeds of Air Force operation. The Naval Research Laboratory continued advancements with their Timation (Time Navigation) satellites, first launched in 1967, and with the third one in 1974 carrying the first atomic clock into orbit.

Another important predecessor to GPS came from a different branch of the United States military. In 1964, the United States Army orbited its first Sequential Collation of Range (SECOR) satellite used for geodetic surveying. The SECOR system included three ground-based transmitters from known locations that would send signals to the satellite transponder in orbit. A fourth ground-based station, at an undetermined position, could then use those signals to fix its location precisely. The last SECOR satellite was launched in 1969.

Decades later, during the early years of GPS, civilian surveying became one of the first fields to make use of the new technology, because surveyors could reap benefits of signals from the less-than-complete GPS constellation years before it was declared operational. GPS can be thought of as an evolution of the SECOR system where the ground-based transmitters have been migrated into orbit.

Development

With these parallel developments in the 1960s, it was realized that a superior system could be developed by synthesizing the best technologies from 621B, Transit, Timation, and SECOR in a multi-service program.

During Labor Day weekend in 1973, a meeting of about twelve military officers at the Pentagon discussed the creation of a *Defense Navigation Satellite System (DNSS)*. It was at this meeting that the real synthesis that became GPS was created. Later that year, the DNSS program was named *Navstar*, or Navigation System Using Timing and Ranging. With the individual satellites being associated with the name Navstar (as with the predecessors Transit and Timation), a more fully encompassing name was used to identify the constellation of Navstar satellites, *Navstar-GPS*. Ten "Block I" prototype satellites were launched between 1978 and 1985 (an additional unit was destroyed in a launch failure).

After Korean Air Lines Flight 007, a Boeing 747 carrying 269 people, was shot down in 1983 after straying into the USSR's prohibited airspace, in the vicinity of Sakhalin and Moneron Islands, President Ronald Reagan issued a directive making GPS freely available for civilian use, once it was sufficiently developed, as a common good. The first Block II satellite was launched on February 14, 1989, and the 24th satellite was launched in 1994. The GPS program cost at this point, not including the cost of the user equipment, but including the costs of the satellite launches, has been estimated at about USD$5 billion (then-year dollars). Roger L. Easton is widely credited as the primary inventor of GPS.

Initially, the highest quality signal was reserved for military use, and the signal available for civilian use was intentionally degraded (Selective Availability). This changed with President Bill Clinton signing a policy directive in 1996 to turn off Selective Availability in May 2000 to provide the same precision to civilians that was afforded to the military. The directive was proposed by the U.S. Secretary of Defense, William Perry, because of the widespread growth of differential GPS services to improve civilian accuracy and eliminate the U.S. military advantage. Moreover, the U.S. military was actively developing technologies to deny GPS service to potential adversaries on a regional basis.

Since its deployment, the U.S. has implemented several improvements to the GPS service including new signals for civil use and increased accuracy and integrity for all users, all the while maintaining compatibility with existing GPS equipment. Modernization of the satellite system has been an ongoing initiative by the U.S. Department of Defense through a series of satellite acquisitions to meet the growing needs of the military, civilians, and the commercial market.

As of early 2015, high-quality, FAA grade, Standard Positioning Service (SPS) GPS receivers provide horizontal accuracy of better than 3.5 meters, although many factors such as receiver quality and atmospheric issues can affect this accuracy.

GPS is owned and operated by the United States Government as a national resource. The Department of Defense is the steward of GPS. *Interagency GPS Executive Board (IGEB)* oversaw GPS policy matters from 1996 to 2004. After that the National Space-Based Positioning, Navigation and Timing Executive Committee was established by presidential directive in 2004 to advise and coordinate federal departments and agencies on matters concerning the GPS and related systems. The executive committee is chaired jointly by the deputy secretaries of defense and transportation. Its membership includes equivalent-level officials from the departments of state, commerce, and homeland security, the joint chiefs of staff, and NASA. Components of the executive office of the president participate as observers to the executive committee, and the FCC chairman participates as a liaison.

The U.S. Department of Defense is required by law to "maintain a Standard Positioning Service (as defined in the federal radio navigation plan and the standard positioning service signal specification) that will be available on a continuous, worldwide basis," and "develop measures to prevent hostile use of GPS and its augmentations without unduly disrupting or degrading civilian uses."

Timeline and Modernization

		\multicolumn{4}{c}{Summary of satellites}	
Block	Launch Period	Satellite launches				Currently in orbit and healthy
		Success	Failure	In preparation	Planned	
I	1978–1985	10	1	0	0	0
II	1989–1990	9	0	0	0	0
IIA	1990–1997	19	0	0	0	0
IIR	1997–2004	12	1	0	0	12
IIR-M	2005–2009	8	0	0	0	7
IIF	2010–2016	12	0	0	0	12
IIIA	From 2017	0	0	0	12	0
IIIB	—	0	0	0	8	0
IIIC	—	0	0	0	16	0
Total		70	2	0	36	31
(Last update: March 9, 2016) 8 satellites from Block IIA are placed in reserve USA-203 from Block IIR-M is unhealthy For a more complete list, see *list of GPS satellite launches*						

- In 1972, the USAF Central Inertial Guidance Test Facility (Holloman AFB), conducted developmental flight tests of two prototype GPS receivers over White Sands Missile Range, using ground-based pseudo-satellites.

- In 1978, the first experimental Block-I GPS satellite was launched.

- In 1983, after Soviet interceptor aircraft shot down the civilian airliner KAL 007 that strayed into prohibited airspace because of navigational errors, killing all 269 people on board, U.S. President Ronald Reagan announced that GPS would be made available for civilian uses once it was completed, although it had been previously published [in Navigation magazine] that the CA code (Coarse Acquisition code) would be available to civilian users.

- By 1985, ten more experimental Block-I satellites had been launched to validate the concept.

- Beginning in 1988, Command & Control of these satellites was transitioned from Onizuka AFS, California to the 2nd Satellite Control Squadron (2SCS) located at Falcon Air Force Station in Colorado Springs, Colorado.

- On February 14, 1989, the first modern Block-II satellite was launched.

- The Gulf War from 1990 to 1991 was the first conflict in which the military widely used GPS.

- In 1991, a project to create a miniature GPS receiver successfully ended, replacing the previous 23 kg military receivers with a 1.25 kg handheld receiver.

- In 1992, the 2nd Space Wing, which originally managed the system, was inactivated and replaced by the 50th Space Wing.

- By December 1993, GPS achieved initial operational capability (IOC), indicating a full constellation (24 satellites) was available and providing the Standard Positioning Service (SPS).

- Full Operational Capability (FOC) was declared by Air Force Space Command (AFSPC) in April 1995, signifying full availability of the military's secure Precise Positioning Service (PPS).

- In 1996, recognizing the importance of GPS to civilian users as well as military users, U.S. President Bill Clinton issued a policy directive declaring GPS a dual-use system and establishing an Interagency GPS Executive Board to manage it as a national asset.

- In 1998, United States Vice President Al Gore announced plans to upgrade GPS with two new civilian signals for enhanced user accuracy and reliability, particularly with respect to aviation safety and in 2000 the United States Congress authorized the effort, referring to it as *GPS III*.

- On May 2, 2000 "Selective Availability" was discontinued as a result of the 1996 executive order, allowing users to receive a non-degraded signal globally.
- In 2004, the United States Government signed an agreement with the European Community establishing cooperation related to GPS and Europe's planned Galileo system.
- In 2004, United States President George W. Bush updated the national policy and replaced the executive board with the National Executive Committee for Space-Based Positioning, Navigation, and Timing.
- November 2004, Qualcomm announced successful tests of assisted GPS for mobile phones.
- In 2005, the first modernized GPS satellite was launched and began transmitting a second civilian signal (L2C) for enhanced user performance.
- On September 14, 2007, the aging mainframe-based Ground Segment Control System was transferred to the new Architecture Evolution Plan.
- On May 19, 2009, the United States Government Accountability Office issued a report warning that some GPS satellites could fail as soon as 2010.
- On May 21, 2009, the Air Force Space Command allayed fears of GPS failure saying "There's only a small risk we will not continue to exceed our performance standard."
- On January 11, 2010, an update of ground control systems caused a software incompatibility with 8000 to 10000 military receivers manufactured by a division of Trimble Navigation Limited of Sunnyvale, Calif.
- On February 25, 2010, the U.S. Air Force awarded the contract to develop the GPS Next Generation Operational Control System (OCX) to improve accuracy and availability of GPS navigation signals, and serve as a critical part of GPS modernization.

Awards

On February 10, 1993, the National Aeronautic Association selected the GPS Team as winners of the 1992 Robert J. Collier Trophy, the nation's most prestigious aviation award. This team combines researchers from the Naval Research Laboratory, the USAF, the Aerospace Corporation, Rockwell International Corporation, and IBM Federal Systems Company. The citation honors them "for the most significant development for safe and efficient navigation and surveillance of air and spacecraft since the introduction of radio navigation 50 years ago."

Two GPS developers received the National Academy of Engineering Charles Stark Draper Prize for 2003:
- Ivan Getting, emeritus president of The Aerospace Corporation and an engineer

at the Massachusetts Institute of Technology, established the basis for GPS, improving on the World War II land-based radio system called LORAN (*Long-range Radio Aid to Navigation*).

- Bradford Parkinson, professor of aeronautics and astronautics at Stanford University, conceived the present satellite-based system in the early 1960s and developed it in conjunction with the U.S. Air Force. Parkinson served twenty-one years in the Air Force, from 1957 to 1978, and retired with the rank of colonel.

GPS developer Roger L. Easton received the National Medal of Technology on February 13, 2006.

Francis X. Kane (Col. USAF, ret.) was inducted into the U.S. Air Force Space and Missile Pioneers Hall of Fame at Lackland A.F.B., San Antonio, Texas, March 2, 2010 for his role in space technology development and the engineering design concept of GPS conducted as part of Project 621B.

In 1998, GPS technology was inducted into the Space Foundation Space Technology Hall of Fame.

On October 4, 2011, the International Astronautical Federation (IAF) awarded the Global Positioning System (GPS) its 60th Anniversary Award, nominated by IAF member, the American Institute for Aeronautics and Astronautics (AIAA). The IAF Honors and Awards Committee recognized the uniqueness of the GPS program and the exemplary role it has played in building international collaboration for the benefit of humanity.

Basic Concept of GPS

Fundamentals

The GPS concept is based on time and the known position of specialized satellites. The satellites carry very stable atomic clocks that are synchronized to each other and to ground clocks. Any drift from true time maintained on the ground is corrected daily. Likewise, the satellite locations are known with great precision. GPS receivers have clocks as well; however, they are not synchronized with true time, and are less stable. GPS satellites continuously transmit their current time and position. A GPS receiver monitors multiple satellites and solves equations to determine the precise position of the receiver and its deviation from true time. At a minimum, four satellites must be in view of the receiver for it to compute four unknown quantities (three position coordinates and clock deviation from satellite time).

More Detailed Description

Each GPS satellite continually broadcasts a signal (carrier wave with modulation) that includes:

- A pseudorandom code (sequence of ones and zeros) that is known to the receiver. By time-aligning a receiver-generated version and the receiver-measured version of the code, the time of arrival (TOA) of a defined point in the code sequence, called an epoch, can be found in the receiver clock time scale.

- A message that includes the time of transmission (TOT) of the code epoch (in GPS system time scale) and the satellite position at that time.

Conceptually, the receiver measures the TOAs (according to its own clock) of four satellite signals. From the TOAs and the TOTs, the receiver forms four time of flight (TOF) values, which are (given the speed of light) approximately equivalent to receiver-satellite range differences. The receiver then computes its three-dimensional position and clock deviation from the four TOFs.

In practice the receiver position (in three dimensional Cartesian coordinates with origin at the Earth's center) and the offset of the receiver clock relative to the GPS time are computed simultaneously, using the navigation equations to process the TOFs.

The receiver's Earth-centered solution location is usually converted to latitude, longitude and height relative to an ellipsoidal Earth model. The height may then be further converted to height relative the geoid (e.g., EGM96) (essentially, mean sea level). These coordinates may be displayed, e.g. on a moving map display and/or recorded and/or used by some other system (e.g., a vehicle guidance system).

User-satellite Geometry

Although usually not formed explicitly in the receiver processing, the conceptual time differences of arrival (TDOAs) define the measurement geometry. Each TDOA corresponds to a hyperboloid of revolution. The line connecting the two satellites involved (and its extensions) forms the axis of the hyperboloid. The receiver is located at the point where three hyperboloids intersect.

It is sometimes incorrectly said that the user location is at the intersection of three spheres. While simpler to visualize, this is only the case if the receiver has a clock synchronized with the satellite clocks (i.e., the receiver measures true ranges to the satellites rather than range differences). There are significant performance benefits to the user carrying a clock synchronized with the satellites. Foremost is that only three satellites are needed to compute a position solution. If this were part of the GPS system concept so that all users needed to carry a synchronized clock, then a smaller number of satellites could be deployed. However, the cost and complexity of the user equipment would increase significantly.

Receiver in Continuous Operation

The description above is representative of a receiver start-up situation. Most receivers have a track algorithm, sometimes called a *tracker*, that combines sets of satellite measurements collected at different times—in effect, taking advantage of the fact that successive receiver positions are usually close to each other. After a set of measurements are processed, the tracker predicts the receiver location corresponding to the next set of satellite measurements. When the new measurements are collected, the receiver uses a weighting scheme to combine the new measurements with the tracker prediction. In general, a tracker can (a) improve receiver position and time accuracy, (b) reject bad measurements, and (c) estimate receiver speed and direction.

The disadvantage of a tracker is that changes in speed or direction can only be computed with a delay, and that derived direction becomes inaccurate when the distance traveled between two position measurements drops below or near the random error of position measurement. GPS units can use measurements of the Doppler shift of the signals received to compute velocity accurately. More advanced navigation systems use additional sensors like a compass or an inertial navigation system to complement GPS.

Non-navigation Applications

In typical GPS operation as a navigator, four or more satellites must be visible to obtain an accurate result. The solution of the navigation equations gives the position of the receiver along with the difference between the time kept by the receiver's on-board clock and the true time-of-day, thereby eliminating the need for a more precise and possibly impractical receiver based clock. Applications for GPS such as time transfer, traffic signal timing, and synchronization of cell phone base stations, make use of this cheap and highly accurate timing. Some GPS applications use this time for display, or, other than for the basic position calculations, do not use it at all.

Although four satellites are required for normal operation, fewer apply in special cases. If one variable is already known, a receiver can determine its position using only three satellites. For example, a ship or aircraft may have known elevation. Some GPS receivers may use additional clues or assumptions such as reusing the last known altitude, dead reckoning, inertial navigation, or including information from the vehicle computer, to give a (possibly degraded) position when fewer than four satellites are visible.

Structure

The current GPS consists of three major segments. These are the space segment (SS), a control segment (CS), and a user segment (US). The U.S. Air Force develops, maintains, and operates the space and control segments. GPS satellites broadcast signals from space, and each GPS receiver uses these signals to calculate its three-dimensional location (latitude, longitude, and altitude) and the current time.

The space segment is composed of 24 to 32 satellites in medium Earth orbit and also includes the payload adapters to the boosters required to launch them into orbit. The control segment is composed of a master control station (MCS), an alternate master control station, and a host of dedicated and shared ground antennas and monitor stations. The user segment is composed of hundreds of thousands of U.S. and allied military users of the secure GPS Precise Positioning Service, and hundreds of millions of civil, commercial, and scientific users of the Standard Positioning Service.

Space Segment

Unlaunched GPS block II-A satellite on display at the San Diego Air & Space Museum.

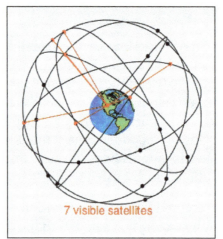

A visual example of a 24 satellite GPS constellation in motion with the earth rotating. Notice how the number of *satellites in view* from a given point on the earth's surface, in this example in Golden CO (39.7469° N, 105.2108° W), changes with time.

The space segment (SS) is composed of the orbiting GPS satellites, or Space Vehicles (SV) in GPS parlance. The GPS design originally called for 24 SVs, eight each in three approximately circular orbits, but this was modified to six orbital planes with four satellites each. The six orbit planes have approximately 55° inclination (tilt relative to the Earth's equator) and are separated by 60° right ascension of the ascending node (angle along the equator from a reference point to the orbit's intersection). The orbital period is one-half a sidereal day, i.e., 11 hours and 58 minutes so that the satellites pass over the same locations or almost the same locations every day. The orbits are arranged so that at least six satellites are always within line of sight from almost everywhere on the Earth's surface. The result of this objective is that the four satellites are not evenly spaced (90 degrees) apart within each orbit. In general terms, the angular difference between satellites in each orbit is 30, 105, 120, and 105 degrees apart, which sum to 360 degrees.

Orbiting at an altitude of approximately 20,200 km (12,600 mi); orbital radius of approximately 26,600 km (16,500 mi), each SV makes two complete orbits each sidereal day, repeating the same ground track each day. This was very helpful during development because even with only four satellites, correct alignment means all four are visible from one spot for a few hours each day. For military operations, the ground track repeat can be used to ensure good coverage in combat zones.

As of February 2016, there are 32 satellites in the GPS constellation, 31 of which are in use. The additional satellites improve the precision of GPS receiver calculations by providing redundant measurements. With the increased number of satellites, the constellation was changed to a nonuniform arrangement. Such an arrangement was shown to improve reliability and availability of the system, relative to a uniform system, when multiple satellites fail. About nine satellites are visible from any point on the ground at any one time, ensuring considerable redundancy over the minimum four satellites needed for a position.

Control Segment

Ground monitor station used from 1984 to 2007, on display at the Air Force Space & Missile Museum.

The control segment is composed of:

1. a master control station (MCS),
2. an alternate master control station,
3. four dedicated ground antennas,
4. six dedicated monitor stations.

The MCS can also access U.S. Air Force Satellite Control Network (AFSCN) ground antennas (for additional command and control capability) and NGA (National Geospatial-Intelligence Agency) monitor stations. The flight paths of the satellites are tracked by dedicated U.S. Air Force monitoring stations in Hawaii, Kwajalein Atoll, Ascension Island, Diego Garcia, Colorado Springs, Colorado and Cape Canaveral, along with shared NGA monitor stations operated in England, Argentina, Ecuador, Bahrain, Australia and Washington DC. The tracking information is sent to the Air Force Space Command MCS at Schriever Air Force Base 25 km (16 mi) ESE of Colorado Springs, which is operated by the 2nd Space Operations Squadron (2 SOPS) of the U.S. Air Force. Then 2 SOPS contacts each GPS satellite regularly with a navigational update using dedicated or shared (AFSCN) ground antennas (GPS dedicated ground antennas are located at Kwajalein, Ascension Island, Diego Garcia, and Cape Canaveral). These updates synchronize the atomic clocks on board the satellites to within a few nanoseconds of each other, and adjust the ephemeris of each satellite's internal orbital model. The updates are created by a Kalman filter that uses inputs from the ground monitoring stations, space weather information, and various other inputs.

Satellite maneuvers are not precise by GPS standards—so to change a satellite's orbit, the satellite must be marked *unhealthy*, so receivers don't use it. After the satellite maneuver, engineers track the new orbit from the ground, upload the new ephemeris, and mark the satellite healthy again.

The Operation Control Segment (OCS) currently serves as the control segment of record. It provides the operational capability that supports GPS users and keeps the GPS system operational and performing within specification.

OCS successfully replaced the legacy 1970s-era mainframe computer at Schriever Air Force Base in September 2007. After installation, the system helped enable upgrades and provide a foundation for a new security architecture that supported U.S. armed forces. OCS will continue to be the ground control system of record until the new segment, Next Generation GPS Operation Control System (OCX), is fully developed and functional.

The new capabilities provided by OCX will be the cornerstone for revolutionizing GPS's mission capabilities, and enabling Air Force Space Command to greatly enhance GPS operational services to U.S. combat forces, civil partners and myriad domestic and international users.

The GPS OCX program also will reduce cost, schedule and technical risk. It is designed to provide 50% sustainment cost savings through efficient software architecture and Performance-Based Logistics. In addition, GPS OCX is expected to cost millions less than the cost to upgrade OCS while providing four times the capability.

The GPS OCX program represents a critical part of GPS modernization and provides significant information assurance improvements over the current GPS OCS program.

- OCX will have the ability to control and manage GPS legacy satellites as well as the next generation of GPS III satellites, while enabling the full array of military signals.

- Built on a flexible architecture that can rapidly adapt to the changing needs of today's and future GPS users allowing immediate access to GPS data and constellation status through secure, accurate and reliable information.

- Provides the warfighter with more secure, actionable and predictive information to enhance situational awareness.

- Enables new modernized signals (L1C, L2C, and L5) and has M-code capability, which the legacy system is unable to do.

- Provides significant information assurance improvements over the current program including detecting and preventing cyber attacks, while isolating, containing and operating during such attacks.

- Supports higher volume near real-time command and control capabilities and abilities.

On September 14, 2011, the U.S. Air Force announced the completion of GPS OCX Preliminary Design Review and confirmed that the OCX program is ready for the next phase of development.

The GPS OCX program has missed major milestones and is pushing the GPS IIIA launch beyond April 2016.

User Segment

GPS receivers come in a variety of formats, from devices integrated into cars, phones, and watches, to dedicated devices such as these.

The first portable GPS unit, Leica WM 101 displayed at the Irish National Science Museum at Maynooth.

The user segment is composed of hundreds of thousands of U.S. and allied military users of the secure GPS Precise Positioning Service, and tens of millions of civil, commercial and scientific users of the Standard Positioning Service. In general, GPS receivers are composed of an antenna, tuned to the frequencies transmitted by the satellites, receiver-processors, and a highly stable clock (often a crystal oscillator). They may also include a display for providing location and speed information to the user. A receiver is often described by its number of channels: this signifies how many satellites it can monitor simultaneously. Originally limited to four or five, this has progressively increased over the years so that, as of 2007, receivers typically have between 12 and 20 channels.

A typical OEM GPS receiver module measuring 15×17 mm.

GPS receivers may include an input for differential corrections, using the RTCM SC-104 format. This is typically in the form of an RS-232 port at 4,800 bit/s speed. Data is actually sent at a much lower rate, which limits the accuracy of the signal sent using RTCM. Receivers with internal DGPS receivers can outperform those using external RTCM data. As of 2006, even low-cost units commonly include Wide Area Augmentation System (WAAS) receivers.

A typical GPS receiver with integrated antenna.

Many GPS receivers can relay position data to a PC or other device using the NMEA 0183 protocol. Although this protocol is officially defined by the National Marine Electronics Association (NMEA), references to this protocol have been compiled from public records, allowing open source tools like gpsd to read the protocol without violating intellectual property laws. Other proprietary protocols exist as well, such as the SiRF and MTK protocols. Receivers can interface with other devices using methods including a serial connection, USB, or Bluetooth.

Applications

While originally a military project, GPS is considered a *dual-use* technology, meaning it has significant military and civilian applications.

GPS has become a widely deployed and useful tool for commerce, scientific uses, tracking, and surveillance. GPS's accurate time facilitates everyday activities such as banking, mobile phone operations, and even the control of power grids by allowing well synchronized hand-off switching.

Civilian

This antenna is mounted on the roof of a hut containing a scientific experiment needing precise timing.

Many civilian applications use one or more of GPS's three basic components: absolute location, relative movement, and time transfer.

- Astronomy: both positional and clock synchronization data is used in astrometry and celestial mechanics calculations. It is also used in amateur astronomy using small telescopes to professionals observatories, for example, while finding extrasolar planets.
- Automated vehicle: applying location and routes for cars and trucks to function without a human driver.
- Cartography: both civilian and military cartographers use GPS extensively.
- Cellular telephony: clock synchronization enables time transfer, which is critical for synchronizing its spreading codes with other base stations to facilitate inter-cell handoff and support hybrid GPS/cellular position detection for mobile emergency calls and other applications. The first handsets with integrated GPS launched in the late 1990s. The U.S. Federal Communications Commission (FCC) mandated the feature in either the handset or in the towers (for use in triangulation) in 2002 so emergency services could locate 911 callers. Third-party

software developers later gained access to GPS APIs from Nextel upon launch, followed by Sprint in 2006, and Verizon soon thereafter.

- Clock synchronization: the accuracy of GPS time signals (±10 ns) is second only to the atomic clocks they are based on.

- Disaster relief/emergency services: depend upon GPS for location and timing capabilities.

- GPS-equipped radiosondes and dropsondes: measure and calculate the atmospheric pressure, wind speed and direction up to 27 km from the Earth's surface

- Radio occultation for weather and atmospheric science applications.

- Fleet tracking: the use of GPS technology to identify, locate and maintain contact reports with one or more fleet vehicles in real-time.

- Geofencing: vehicle tracking systems, person tracking systems, and pet tracking systems use GPS to locate a vehicle, person, or pet. These devices are attached to the vehicle, person, or the pet collar. The application provides continuous tracking and mobile or Internet updates should the target leave a designated area.

- Geotagging: applying location coordinates to digital objects such as photographs (in Exif data) and other documents for purposes such as creating map overlays with devices like Nikon GP-1

- GPS aircraft tracking

- GPS for mining: the use of RTK GPS has significantly improved several mining operations such as drilling, shoveling, vehicle tracking, and surveying. RTK GPS provides centimeter-level positioning accuracy.

- GPS data mining: It is possible to use GPS data from multiple users to understand movement patterns. It is possible to aggregate data from multiple users to understand common trajectories and interesting locations.

- GPS tours: location determines what content to display; for instance, information about an approaching point of interest.

- Navigation: navigators value digitally precise velocity and orientation measurements.

- Phasor measurements: GPS enables highly accurate timestamping of power system measurements, making it possible to compute phasors.

- Recreation: for example, geocaching, geodashing, GPS drawing and waymarking.

- Robotics: self-navigating, autonomous robots using a GPS sensors, which calculate latitude, longitude, time, speed, and heading.

- Sport: used in football and rugby for the control and analysis of the training load.

- Surveying: surveyors use absolute locations to make maps and determine property boundaries.

- Tectonics: GPS enables direct fault motion measurement of earthquakes. Between earthquakes GPS can be used to measure crustal motion and deformation to estimate seismic strain buildup for creating seismic hazard maps.

- Telematics: GPS technology integrated with computers and mobile communications technology in automotive navigation systems.

Restrictions on Civilian Use

The U.S. government controls the export of some civilian receivers. All GPS receivers capable of functioning above 18 km (60,000 feet) altitude and 515 m/s (1,000 knots), or designed or modified for use with unmanned air vehicles like, e.g., ballistic or cruise missile systems, are classified as munitions (weapons)—which means they require State Department export licenses.

This rule applies even to otherwise purely civilian units that only receive the L1 frequency and the C/A (Coarse/Acquisition) code.

Disabling operation above these limits exempts the receiver from classification as a munition. Vendor interpretations differ. The rule refers to operation at both the target altitude and speed, but some receivers stop operating even when stationary. This has caused problems with some amateur radio balloon launches that regularly reach 30 km (100,000 feet).

These limits only apply to units or components exported from the USA. A growing trade in various components exists, including GPS units from other countries. These are expressly sold as ITAR-free.

Military

Attaching a GPS guidance kit to a dumb bomb, March 2003.

As of 2009, military GPS applications include:

- Navigation: Soldiers use GPS to find objectives, even in the dark or in unfamiliar territory, and to coordinate troop and supply movement. In the United States armed forces, commanders use the *Commanders Digital Assistant* and lower ranks use the *Soldier Digital Assistant*.

- Target tracking: Various military weapons systems use GPS to track potential ground and air targets before flagging them as hostile. These weapon systems pass target coordinates to precision-guided munitions to allow them to engage targets accurately. Military aircraft, particularly in air-to-ground roles, use GPS to find targets.

- Missile and projectile guidance: GPS allows accurate targeting of various military weapons including ICBMs, cruise missiles, precision-guided munitions and Artillery projectiles. Embedded GPS receivers able to withstand accelerations of 12,000 g or about 118 km/s² have been developed for use in 155-millimeter (6.1 in) howitzer shells.

- Search and rescue.

- Reconnaissance: Patrol movement can be managed more closely.

- GPS satellites carry a set of nuclear detonation detectors consisting of an optical sensor (Y-sensor), an X-ray sensor, a dosimeter, and an electromagnetic pulse (EMP) sensor (W-sensor), that form a major portion of the United States Nuclear Detonation Detection System. General William Shelton has stated that future satellites may drop this feature to save money.

GPS type navigation was first used in war in the 1991 Persian Gulf War, before GPS was fully developed in 1995, to assist Coalition Forces to navigate and perform maneuvers in the war. The war also demonstrated the vulnerability of GPS to being jammed, when Iraqi forces added noise to the weak GPS signal transmission to protect Iraqi targets.

M982 Excalibur GPS-guided artillery shell.

Communication

The navigational signals transmitted by GPS satellites encode a variety of information including satellite positions, the state of the internal clocks, and the health of the network. These signals are transmitted on two separate carrier frequencies that are common to all satellites in the network. Two different encodings are used: a public encoding that enables lower resolution navigation, and an encrypted encoding used by the U.S. military.

Message Format

GPS message format	
Subframes	**Description**
1	Satellite clock, GPS time relationship
2–3	Ephemeris (precise satellite orbit)
4–5	Almanac component (satellite network synopsis, error correction)

Each GPS satellite continuously broadcasts a *navigation message* on L1 C/A and L2 P/Y frequencies at a rate of 50 bits per second. Each complete message takes 750 seconds (12 1/2 minutes) to complete. The message structure has a basic format of a 1500-bit-long frame made up of five subframes, each subframe being 300 bits (6 seconds) long. Subframes 4 and 5 are subcommutated 25 times each, so that a complete data message requires the transmission of 25 full frames. Each subframe consists of ten words, each 30 bits long. Thus, with 300 bits in a subframe times 5 subframes in a frame times 25 frames in a message, each message is 37,500 bits long. At a transmission rate of 50-bit/s, this gives 750 seconds to transmit an entire almanac message (GPS). Each 30-second frame begins precisely on the minute or half-minute as indicated by the atomic clock on each satellite.

The first subframe of each frame encodes the week number and the time within the week, as well as the data about the health of the satellite. The second and the third subframes contain the *ephemeris* – the precise orbit for the satellite. The fourth and fifth subframes contain the *almanac*, which contains coarse orbit and status information for up to 32 satellites in the constellation as well as data related to error correction. Thus, to obtain an accurate satellite location from this transmitted message, the receiver must demodulate the message from each satellite it includes in its solution for 18 to 30 seconds. To collect all transmitted almanacs, the receiver must demodulate the message for 732 to 750 seconds or 12 1/2 minutes.

All satellites broadcast at the same frequencies, encoding signals using unique code division multiple access (CDMA) so receivers can distinguish individual satellites from each other. The system uses two distinct CDMA encoding types: the coarse/acquisition (C/A) code, which is accessible by the general public, and the precise (P(Y)) code, which is encrypted so that only the U.S. military and other NATO nations who have been given access to the encryption code can access it.

The ephemeris is updated every 2 hours and is generally valid for 4 hours, with provisions for updates every 6 hours or longer in non-nominal conditions. The almanac is updated typically every 24 hours. Additionally, data for a few weeks following is uploaded in case of transmission updates that delay data upload.

Satellite Frequencies

GPS frequency overview		
Band	**Frequency**	**Description**
L1	1575.42 MHz	Coarse-acquisition (C/A) and encrypted precision (P(Y)) codes, plus the L1 civilian (L1C) and military (M) codes on future Block III satellites.
L2	1227.60 MHz	P(Y) code, plus the L2C and military codes on the Block IIR-M and newer satellites.
L3	1381.05 MHz	Used for nuclear detonation (NUDET) detection.
L4	1379.913 MHz	Being studied for additional ionospheric correction.
L5	1176.45 MHz	Proposed for use as a civilian safety-of-life (SoL) signal.

All satellites broadcast at the same two frequencies, 1.57542 GHz (L1 signal) and 1.2276 GHz (L2 signal). The satellite network uses a CDMA spread-spectrum technique where the low-bitrate message data is encoded with a high-rate pseudo-random (PRN) sequence that is different for each satellite. The receiver must be aware of the PRN codes for each satellite to reconstruct the actual message data. The C/A code, for civilian use, transmits data at 1.023 million chips per second, whereas the P code, for U.S. military use, transmits at 10.23 million chips per second. The actual internal reference of the satellites is 10.22999999543 MHz to compensate for relativistic effects that make observers on the Earth perceive a different time reference with respect to the transmitters in orbit. The L1 carrier is modulated by both the C/A and P codes, while the L2 carrier is only modulated by the P code. The P code can be encrypted as a so-called P(Y) code that is only available to military equipment with a proper decryption key. Both the C/A and P(Y) codes impart the precise time-of-day to the user.

The L3 signal at a frequency of 1.38105 GHz is used to transmit data from the satellites to ground stations. This data is used by the United States Nuclear Detonation (NUDET) Detection System (USNDS) to detect, locate, and report nuclear detonations (NUDETs) in the Earth's atmosphere and near space. One usage is the enforcement of nuclear test ban treaties.

The L4 band at 1.379913 GHz is being studied for additional ionospheric correction.

The L5 frequency band at 1.17645 GHz was added in the process of GPS modernization. This frequency falls into an internationally protected range for aeronautical navigation, promising little or no interference under all circumstances. The first Block IIF satellite that provides this signal was launched in 2010. The L5 consists of two carrier components that are in phase quadrature with each other. Each carrier component is bi-phase shift key (BPSK) modulated by a separate bit train. "L5, the third civil GPS signal, will eventually support safety-of-life applications for aviation and provide improved availability and accuracy."

A conditional waiver has recently (2011-01-26) been granted to LightSquared to operate a terrestrial broadband service near the L1 band. Although LightSquared had applied for a license to operate in the 1525 to 1559 band as early as 2003 and it was put out for public comment, the FCC asked LightSquared to form a study group with the GPS community to test GPS receivers and identify issue that might arise due to the larger signal power from the LightSquared terrestrial network. The GPS community had not objected to the LightSquared (formerly MSV and SkyTerra) applications until November 2010, when LightSquared applied for a modification to its Ancillary Terrestrial Component (ATC) authorization. This filing (SAT-MOD-20101118-00239) amounted to a request to run several orders of magnitude more power in the same frequency band for terrestrial base stations, essentially repurposing what was supposed to be a "quiet neighborhood" for signals from space as the equivalent of a cellular network. Testing in the first half of 2011 has demonstrated that the impact of the lower 10 MHz of spectrum is minimal to GPS devices (less than 1% of the total GPS devices are affected). The upper 10 MHz intended for use by LightSquared may have some impact on GPS devices. There is some concern that this may seriously degrade the GPS signal for many consumer uses. Aviation Week magazine reports that the latest testing (June 2011) confirms "significant jamming" of GPS by LightSquared's system.

Demodulation and Decoding

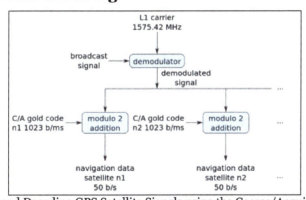

Demodulating and Decoding GPS Satellite Signals using the Coarse/Acquisition Gold code.

Because all of the satellite signals are modulated onto the same L1 carrier frequency, the signals must be separated after demodulation. This is done by assigning each satellite a unique binary sequence known as a Gold code. The signals are decoded after demodulation using addition of the Gold codes corresponding to the satellites monitored by the receiver.

If the almanac information has previously been acquired, the receiver picks the satellites to listen for by their PRNs, unique numbers in the range 1 through 32. If the almanac information is not in memory, the receiver enters a search mode until a lock is obtained on one of the satellites. To obtain a lock, it is necessary that there be an unobstructed line of sight from the receiver to the satellite. The receiver can then acquire the almanac and determine the satellites it should listen for. As it detects each satellite's signal, it identifies it by its distinct C/A code pattern. There can be a delay of up to 30 seconds before the first estimate of position because of the need to read the ephemeris data.

Processing of the navigation message enables the determination of the time of transmission and the satellite position at this time.

Navigation Equations

Problem Description

The receiver uses messages received from satellites to determine the satellite positions and time sent. The x, y, and z components of satellite position and the time sent are designated as $[x_i, y_i, z_i, s_i]$ where the subscript i denotes the satellite and has the value 1, 2, ..., n, where $n \geq 4$. When the time of message reception indicated by the on-board receiver clock is $\tilde{t}_i$, the true reception time is $t_i = \tilde{t}_i - b$, where b is the receiver's clock bias from the much more accurate GPS system clocks employed by the satellites. The receiver clock bias is the same for all received satellite signals (assuming the satellite clocks are all perfectly synchronized). The message's transit time is $\tilde{t}_i - b - s_i$, where s_i is the satellite time. Assuming the message traveled at the speed of light, c, the distance traveled is $(\tilde{t}_i - b - s_i)\,c$.

For n satellites, the equations to satisfy are:

$$(x-x_i)^2 + (y-y_i)^2 + (z-z_i)^2 = \left([\tilde{t}_i - b - s_i]c\right)^2, i = 1, 2, \ldots, n$$

or in terms of *pseudoranges*, $p_i = (\tilde{t}_i - s_i)c$, as:

$$\sqrt{(x-x_i)^2 + (y-y_i)^2 + (z-z_i)^2} + bc = p_i, i = 1, 2, \ldots, n.$$

Since the equations have four unknowns [x, y, z, b]—the three components of GPS receiver position and the clock bias—signals from at least four satellites are necessary to attempt solving these equations. They can be solved by algebraic or numerical methods. Existence and uniqueness of GPS solutions are discussed by Abell and Chaffee. When n is greater than 4 this system is overdetermined and a fitting method must be used.

With each combination of satellites, GDOP quantities can be calculated based on the relative sky directions of the satellites used. The receiver location is expressed in a specific coordinate system, such as latitude and longitude using the WGS 84 geodetic datum or a country-specific system.

Geometric Interpretation

The GPS equations can be solved by numerical and analytical methods. Geometrical interpretations can enhance the understanding of these solution methods.

Spheres

The measured ranges, called pseudoranges, contain clock errors. In a simplified idealization in which the ranges are synchronized, these true ranges represent the radii of spheres, each centered on one of the transmitting satellites. The solution for the position of the receiver is then at the intersection of the surfaces of three of these spheres. If more than the minimum number of ranges is available, a near intersection of more than three sphere surfaces could be found via, e.g. least squares.

Hyperboloids

If the distance traveled between the receiver and satellite i and the distance traveled between the receiver and satellite j are subtracted, the result is $(\tilde{t}_i - s_i) c - (\tilde{t}_j - s_j) c$, which only involves known or measured quantities. The locus of points having a constant difference in distance to two points (here, two satellites) is a hyperboloid. Thus, from four or more measured reception times, the receiver can be placed at the intersection of the surfaces of three or more hyperboloids.

Spherical Cones

The solution space [x, y, z, b] can be seen as a four-dimensional geometric space. In that case each of the equations describes a spherical cone, with the cusp located at the satellite, and the base a sphere around the satellite. The receiver is at the intersection of four or more of such cones.

Solution Methods

Least Squares

When more than four satellites are available, the calculation can use the four best, or more than four simultaneously (up to all visible satellites), depending on the number of receiver channels, processing capability, and geometric dilution of precision (GDOP).

Using more than four involves an over-determined system of equations with no unique solution; such a system can be solved by a least-squares or weighted least squares method.

$$\left(\hat{x}, \hat{y}, \hat{z}, \hat{b}\right) = \underset{(x,y,z,b)}{\arg\min} \sum_i \left(\sqrt{(x-x_i)^2 + (y-y_i)^2 + (z-z_i)^2} + bc - p_i\right)^2$$

Iterative

Both the equations for four satellites, or the least squares equations for more than four, are non-linear and need special solution methods. A common approach is by iteration on a linearized form of the equations, (e.g., Gauss–Newton algorithm).

The GPS system was initially developed assuming use of a numerical least-squares solution method—i.e., before closed-form solutions were found.

Closed-form

One closed-form solution to the above set of equations was developed by S. Bancroft. Its properties are well known; in particular, proponents claim it is superior in low-GDOP situations, compared to iterative least squares methods.

Bancroft's method is algebraic, as opposed to numerical, and can be used for four or more satellites. When four satellites are used, the key steps are inversion of a 4x4 matrix and solution of a single-variable quadratic equation. Bancroft's method provides one or two solutions for the unknown quantities. When there are two (usually the case), only one is a near-Earth sensible solution.

When a receiver uses more than four satellites for a solution, Bancroft uses the generalized inverse (i.e., the pseudoinverse) to find a solution. However, a case has been made that iterative methods (e.g., Gauss–Newton algorithm) for solving over-determined non-linear least squares (NLLS) problems generally provide more accurate solutions.

Leick et al. (2015) states that "Bancroft's (1985) solution is a very early, if not the first, closed-form solution." Other closed-form solutions were published afterwards, although their adoption in practice is unclear.

Error Sources and Analysis

GPS error analysis examines error sources in GPS results and the expected size of those errors. GPS makes corrections for receiver clock errors and other effects, but some residual errors remain uncorrected. Error sources include signal arrival time measurements, numerical calculations, atmospheric effects (ionospheric/tropospheric delays), ephemeris and clock data, multipath signals, and natural and artificial interference. Magnitude of residual errors from these sources depends on geometric dilution of precision. Artificial errors may result from jamming devices and threaten ships and aircraft or from intentional signal degradation through selective availability, which limited accuracy to ~6–12 m, but has been switched off since May 1, 2000.

Accuracy Enhancement and Surveying

Augmentation

Integrating external information into the calculation process can materially improve accuracy. Such augmentation systems are generally named or described based on how the information arrives. Some systems transmit additional error information (such as clock drift, ephemera, or ionospheric delay), others characterize prior errors, while a third group provides additional navigational or vehicle information.

Examples of augmentation systems include the Wide Area Augmentation System (WAAS), European Geostationary Navigation Overlay Service (EGNOS), Differential GPS (DGPS), inertial navigation systems (INS) and Assisted GPS. The standard accuracy of about 15 meters (49 feet) can be augmented to 3–5 meters (9.8–16.4 ft) with DGPS, and to about 3 meters (9.8 feet) with WAAS.

Precise Monitoring

Accuracy can be improved through precise monitoring and measurement of existing GPS signals in additional or alternate ways.

The largest remaining error is usually the unpredictable delay through the ionosphere. The spacecraft broadcast ionospheric model parameters, but some errors remain. This is one reason GPS spacecraft transmit on at least two frequencies, L1 and L2. Ionospheric delay is a well-defined function of frequency and the total electron content (TEC) along the path, so measuring the arrival time difference between the frequencies determines TEC and thus the precise ionospheric delay at each frequency.

Military receivers can decode the P(Y) code transmitted on both L1 and L2. Without decryption keys, it is still possible to use a *codeless* technique to compare the P(Y) codes on L1 and L2 to gain much of the same error information. However, this technique is slow, so it is currently available only on specialized surveying equipment. In the future, additional civilian codes are expected to be transmitted on the L2 and L5 frequencies. All users will then be able to perform dual-frequency measurements and directly compute ionospheric delay errors.

A second form of precise monitoring is called *Carrier-Phase Enhancement* (CPGPS). This corrects the error that arises because the pulse transition of the PRN is not instantaneous, and thus the correlation (satellite–receiver sequence matching) operation is imperfect. CPGPS uses the L1 carrier wave, which has a period of $\frac{1s}{1575.42 \times 10^6} = 0.63475 \text{ns} \approx 1\text{ns}$, which is about one-thousandth of the C/A Gold code bit period of $\frac{1s}{1023 \times 10^3} = 977.5 \text{ns} \approx 1000 \text{ns}$, to act as an additional clock signal and resolve the uncertainty. The phase difference error in the normal GPS amounts to 2–3 meters (7–10 ft) of ambiguity. CPGPS working to within 1% of perfect transition reduces this error to 3 centimeters (1.2 in) of ambiguity. By eliminating this error source, CPGPS coupled with DGPS normally realizes between 20–30 centimeters (8–12 in) of absolute accuracy.

Relative Kinematic Positioning (RKP) is a third alternative for a precise GPS-based positioning system. In this approach, determination of range signal can be resolved to a precision of less than 10 centimeters (4 in). This is done by resolving the number of cycles that the signal is transmitted and received by the receiver by using a combination of differential GPS (DGPS) correction data, transmitting GPS signal phase information and ambiguity resolution techniques via statistical tests—possibly with processing in real-time (real-time kinematic positioning, RTK).

Timekeeping

Leap Seconds

While most clocks derive their time from Coordinated Universal Time (UTC), the atomic clocks on the satellites are set to GPS time. The difference is that GPS time is not corrected to match the rotation of the Earth, so it does not contain leap seconds or other corrections that are periodically added to UTC. GPS time was set to match UTC in 1980, but has since diverged. The lack of corrections means that GPS time remains at a constant offset with International Atomic Time (TAI) (TAI − GPS = 19 seconds). Periodic corrections are performed to the on-board clocks to keep them synchronized with ground clocks.

The GPS navigation message includes the difference between GPS time and UTC. As of July 2015, GPS time is 17 seconds ahead of UTC because of the leap second added to UTC on June 30, 2015. Receivers subtract this offset from GPS time to calculate UTC and specific timezone values. New GPS units may not show the correct UTC time until after receiving the UTC offset message. The GPS-UTC offset field can accommodate 255 leap seconds (eight bits).

Accuracy

GPS time is theoretically accurate to about 14 nanoseconds. However, most receivers lose accuracy in the interpretation of the signals and are only accurate to 100 nanoseconds.

Format

As opposed to the year, month, and day format of the Gregorian calendar, the GPS date is expressed as a week number and a seconds-into-week number. The week number is transmitted as a ten-bit field in the C/A and P(Y) navigation messages, and so it becomes zero again every 1,024 weeks (19.6 years). GPS week zero started at 00:00:00 UTC (00:00:19 TAI) on January 6, 1980, and the week number became zero again for the first time at 23:59:47 UTC on August 21, 1999 (00:00:19 TAI on August 22, 1999). To determine the current Gregorian date, a GPS receiver must be provided with the approximate date (to within 3,584 days) to correctly translate the GPS date signal. To address this concern the modernized GPS navigation message uses a 13-bit field that only repeats every 8,192 weeks (157 years), thus lasting until the year 2137 (157 years after GPS week zero).

Carrier Phase Tracking (Surveying)

Another method that is used in surveying applications is carrier phase tracking. The period of the carrier frequency multiplied by the speed of light gives the wavelength, which is about 0.19 meters for the L1 carrier. Accuracy within 1% of wavelength in detecting the leading edge reduces this component of pseudorange error to as little as 2 millimeters. This compares to 3 meters for the C/A code and 0.3 meters for the P code.

However, 2 millimeter accuracy requires measuring the total phase—the number of waves multiplied by the wavelength plus the fractional wavelength, which requires specially equipped receivers. This method has many surveying applications. It is accurate enough for real-time tracking of the very slow motions of tectonic plates, typically 0–100 mm (0–4 inches) per year.

Triple differencing followed by numerical root finding, and a mathematical technique called least squares can estimate the position of one receiver given the position of another. First, compute the difference between satellites, then between receivers, and finally between epochs. Other orders of taking differences are equally valid. Detailed discussion of the errors is omitted.

The satellite carrier total phase can be measured with ambiguity as to the number of cycles. Let $\phi(r_i, s_j, t_k)$ denote the phase of the carrier of satellite j measured by receiver i at time t_k. This notation shows the meaning of the subscripts i, j, and k. The receiver (r), satellite (s), and time (t) come in alphabetical order as arguments of ϕ and to balance readability and conciseness, let $\phi_{i,j,k} = \phi(r_i, s_j, t_k)$ be a concise abbreviation. Also we define three functions,: $\Delta^r, \Delta^s, \Delta^t$, which return differences between receivers, satellites, and time points, respectively. Each function has variables with three subscripts as its arguments. These three functions are defined below. If $\alpha_{i,j,k}$ is a function of the three integer arguments, i, j, and k then it is a valid argument for the functions, : $\Delta^r, \Delta^s, \Delta^t$, with the values defined as:

$$\Delta^r(\alpha_{i,j,k}) = \alpha_{i+1,j,k} - \alpha_{i,j,k},$$

$$\Delta^s(\alpha_{i,j,k}) = \alpha_{i,j+1,k} - \alpha_{i,j,k}, \text{ and}$$

$$\Delta^t(\alpha_{i,j,k}) = \alpha_{i,j,k+1} - \alpha_{i,j,k}.$$

Also if $\alpha_{i,j,k}$ and $\beta_{l,m,n}$ are valid arguments for the three functions and a and b are constants then $(a\,\alpha_{i,j,k} + b\,\beta_{l,m,n})$ is a valid argument with values defined as,

$$\Delta^r(a\,\alpha_{i,j,k}+b\,\beta_{l,m,n})=a\,\Delta^r(\alpha_{i,j,k})+b\,\Delta^r(\beta_{l,m,n}),$$

$$\Delta^s(a\,\alpha_{i,j,k}+b\,\beta_{l,m,n})=a\,\Delta^s(\alpha_{i,j,k})+b\,\Delta^s(\beta_{l,m,n}),$$

$$\Delta^t(a\,\alpha_{i,j,k}+b\,\beta_{l,m,n})=a\,\Delta^t(\alpha_{i,j,k})+b\,\Delta^t(\beta_{l,m,n}).$$

Receiver clock errors can be approximately eliminated by differencing the phases measured from satellite 1 with that from satellite 2 at the same epoch. This difference is designated as $\Delta^s(\phi_{1,1,1})=\phi_{1,2,1}-\phi_{1,1,1}$

Double differencing computes the difference of receiver 1's satellite difference from that of receiver 2. This approximately eliminates satellite clock errors. This double difference is:

$$\Delta^r(\Delta^s(\phi_{1,1,1}))=\Delta^r(\phi_{1,2,1}-\phi_{1,1,1})=\Delta^r(\phi_{1,2,1})-\Delta^r(\phi_{1,1,1})=(\phi_{2,2,1}-\phi_{1,2,1})-(\phi_{2,1,1}-\phi_{1,1,1})$$

Triple differencing subtracts the receiver difference from time 1 from that of time 2. This eliminates the ambiguity associated with the integral number of wavelengths in carrier phase provided this ambiguity does not change with time. Thus the triple difference result eliminates practically all clock bias errors and the integer ambiguity. Atmospheric delay and satellite ephemeris errors have been significantly reduced. This triple difference is:

$$\Delta^t(\Delta^r(\Delta^s(\phi_{1,1,1})))$$

Triple difference results can be used to estimate unknown variables. For example, if the position of receiver 1 is known but the position of receiver 2 unknown, it may be possible to estimate the position of receiver 2 using numerical root finding and least squares. Triple difference results for three independent time pairs may be sufficient to solve for receiver 2's three position components. This may require a numerical procedure. An approximation of receiver 2's position is required to use such a numerical method. This initial value can probably be provided from the navigation message and the intersection of sphere surfaces. Such a reasonable estimate can be key to successful multidimensional root finding. Iterating from three time pairs and a fairly good initial value produces one observed triple difference result for receiver 2's position. Processing additional time pairs can improve accuracy, overdetermining the answer with multiple solutions. Least squares can estimate an overdetermined system. Least squares determines the position of receiver 2 that best fits the observed triple difference results for receiver 2 positions under the criterion of minimizing the sum of the squares.

Regulatory Spectrum Issues Concerning GPS Receivers

In the United States, GPS receivers are regulated under the Federal Communications

Commission's (FCC) Part 15 rules. As indicated in the manuals of GPS-enabled devices sold in the United States, as a Part 15 device, it "must accept any interference received, including interference that may cause undesired operation." With respect to GPS devices in particular, the FCC states that GPS receiver manufacturers, "must use receivers that reasonably discriminate against reception of signals outside their allocated spectrum." For the last 30 years, GPS receivers have operated next to the Mobile Satellite Service band, and have discriminated against reception of mobile satellite services, such as Inmarsat, without any issue.

The spectrum allocated for GPS L1 use by the FCC is 1559 to 1610 MHz, while the spectrum allocated for satellite-to-ground use owned by Lightsquared is the Mobile Satellite Service band. Since 1996, the FCC has authorized licensed use of the spectrum neighboring the GPS band of 1525 to 1559 MHz to the Virginia company LightSquared. On March 1, 2001, the FCC received an application from LightSquared's predecessor, Motient Services to use their allocated frequencies for an integrated satellite-terrestrial service. In 2002, the U.S. GPS Industry Council came to an out-of-band-emissions (OOBE) agreement with LightSquared to prevent transmissions from LightSquared's ground-based stations from emitting transmissions into the neighboring GPS band of 1559 to 1610 MHz. In 2004, the FCC adopted the OOBE agreement in its authorization for LightSquared to deploy a ground-based network ancillary to their satellite system – known as the Ancillary Tower Components (ATCs) – "We will authorize MSS ATC subject to conditions that ensure that the added terrestrial component remains ancillary to the principal MSS offering. We do not intend, nor will we permit, the terrestrial component to become a stand-alone service." This authorization was reviewed and approved by the U.S. Interdepartment Radio Advisory Committee, which includes the U.S. Department of Agriculture, U.S. Air Force, U.S. Army, U.S. Coast Guard, Federal Aviation Administration, National Aeronautics and Space Administration, Interior, and U.S. Department of Transportation.

In January 2011, the FCC conditionally authorized LightSquared's wholesale customers—such as Best Buy, Sharp, and C Spire—to only purchase an integrated satellite-ground-based service from LightSquared and re-sell that integrated service on devices that are equipped to only use the ground-based signal using LightSquared's allocated frequencies of 1525 to 1559 MHz. In December 2010, GPS receiver manufacturers expressed concerns to the FCC that LightSquared's signal would interfere with GPS receiver devices although the FCC's policy considerations leading up to the January 2011 order did not pertain to any proposed changes to the maximum number of ground-based LightSquared stations or the maximum power at which these stations could operate. The January 2011 order makes final authorization contingent upon studies of GPS interference issues carried out by a LightSquared led working group along with GPS industry and Federal agency participation. On February 14, 2012, the FCC initiated proceedings to vacate LightSquared's Conditional Waiver Order based on the NTIA's conclusion that there was currently no practical way to mitigate potential GPS interference.

GPS receiver manufacturers design GPS receivers to use spectrum beyond the GPS-allocated band. In some cases, GPS receivers are designed to use up to 400 MHz of spectrum in either direction of the L1 frequency of 1575.42 MHz, because mobile satellite services in those regions are broadcasting from space to ground, and at power levels commensurate with mobile satellite services. However, as regulated under the FCC's Part 15 rules, GPS receivers are not warranted protection from signals outside GPS-allocated spectrum. This is why GPS operates next to the Mobile Satellite Service band, and also why the Mobile Satellite Service band operates next to GPS. The symbiotic relationship of spectrum allocation ensures that users of both bands are able to operate cooperatively and freely.

The FCC adopted rules in February 2003 that allowed Mobile Satellite Service (MSS) licensees such as LightSquared to construct a small number of ancillary ground-based towers in their licensed spectrum to "promote more efficient use of terrestrial wireless spectrum." In those 2003 rules, the FCC stated "As a preliminary matter, terrestrial [Commercial Mobile Radio Service ("CMRS")] and MSS ATC are expected to have different prices, coverage, product acceptance and distribution; therefore, the two services appear, at best, to be imperfect substitutes for one another that would be operating in predominately different market segments... MSS ATC is unlikely to compete directly with terrestrial CMRS for the same customer base...". In 2004, the FCC clarified that the ground-based towers would be ancillary, noting that "We will authorize MSS ATC subject to conditions that ensure that the added terrestrial component remains ancillary to the principal MSS offering. We do not intend, nor will we permit, the terrestrial component to become a stand-alone service." In July 2010, the FCC stated that it expected LightSquared to use its authority to offer an integrated satellite-terrestrial service to "provide mobile broadband services similar to those provided by terrestrial mobile providers and enhance competition in the mobile broadband sector." However, GPS receiver manufacturers have argued that LightSquared's licensed spectrum of 1525 to 1559 MHz was never envisioned as being used for high-speed wireless broadband based on the 2003 and 2004 FCC ATC rulings making clear that the Ancillary Tower Component (ATC) would be, in fact, ancillary to the primary satellite component. To build public support of efforts to continue the 2004 FCC authorization of LightSquared's ancillary terrestrial component vs. a simple ground-based LTE service in the Mobile Satellite Service band, GPS receiver manufacturer Trimble Navigation Ltd. formed the "Coalition To Save Our GPS."

The FCC and LightSquared have each made public commitments to solve the GPS interference issue before the network is allowed to operate. However, according to Chris Dancy of the Aircraft Owners and Pilots Association, airline pilots with the type of systems that would be affected "may go off course and not even realize it." The problems could also affect the Federal Aviation Administration upgrade to the air traffic control system, United States Defense Department guidance, and local emergency services including 911.

On February 14, 2012, the U.S. Federal Communications Commission (FCC) moved to bar LightSquared's planned national broadband network after being informed by the National Telecommunications and Information Administration (NTIA), the federal agency that coordinates spectrum uses for the military and other federal government entities, that "there is no practical way to mitigate potential interference at this time". LightSquared is challenging the FCC's action.

Other Systems

Comparison of geostationary, GPS, GLONASS, Galileo, Compass (MEO), International Space Station, Hubble Space Telescope and Iridium constellation orbits, with the Van Allen radiation belts and the Earth to scale.[b] The Moon's orbit is around 9 times larger than geostationary orbit.[c] (In the SVG file, hover over an orbit or its label to highlight it; click to load its article.)

Other satellite navigation systems in use or various states of development include:

- GLONASS – Russia's global navigation system. Fully operational worldwide.

- Galileo – a global system being developed by the European Union and other partner countries, planned to be operational by 2016 (and fully deployed by 2020).

- Beidou – People's Republic of China's regional system, currently limited to Asia and the West Pacific, global coverage planned to be operational by 2020.

- IRNSS (NAVIC) – India's regional navigation system, covering India and Northern Indian Ocean.

- QZSS – Japanese regional system covering Asia and Oceania.

Google Earth

Google Earth is a virtual globe, map and geographical information program that was originally called EarthViewer 3D created by Keyhole, Inc, a Central Intelligence Agency (CIA) funded company acquired by Google in 2004. It maps the Earth by the superimposition of images obtained from satellite imagery, aerial photography and geographic information system (GIS) onto a 3D globe. It was originally available with three different licenses, but has since been reduced to just two: Google Earth (a free version with limited function) and Google Earth Pro, which is now free (it previously cost $399 a year) and is intended for commercial use. The third original option, Google Earth Plus, has been discontinued.

The product, re-released as Google Earth in 2005, is available for use on personal computers running Windows 2000 and above, Mac OS X 10.3.9 and above, Linux kernel: 2.6 or later (released on June 12, 2006), and FreeBSD. Google Earth is also available as a browser plugin which was released on May 28, 2008. It was also made available for mobile viewers on the iPhone OS on October 28, 2008, as a free download from the App Store, and is available to Android users as a free app in the Google Play store. In addition to releasing an updated Keyhole based client, Google also added the imagery from the Earth database to their web-based mapping software, Google Maps. The release of Google Earth in June 2005 to the public caused a more than tenfold increase in media coverage on virtual globes between 2004 and 2005, driving public interest in geospatial technologies and applications. As of October 2011, Google Earth has been downloaded more than a billion times.

Google Earth displays satellite images of varying resolution of the Earth's surface, allowing users to see things like cities and houses looking perpendicularly down or at an oblique angle. The degree of resolution available is based somewhat on the points of interest and popularity, but most land (except for some islands) is covered in at least 15 meters of resolution. Maps showing a visual representation of Google Earth coverage Melbourne, Victoria, Australia; Las Vegas, Nevada, USA; and Cambridge, Cambridgeshire, United Kingdom include examples of the highest resolution, at 15 cm (6 inches). Google Earth allows users to search for addresses for some countries, enter coordinates, or simply use the mouse to browse to a location.

For large parts of the surface of the Earth only 2D images are available, from almost vertical photography. Viewing this from an oblique angle, there is perspective in the sense that objects which are horizontally far away are seen smaller, like viewing a large photograph, not quite like a 3D view.

For other parts of the surface of the Earth, 3D images of terrain and buildings are available. Google Earth uses digital elevation model (DEM) data collected by NASA's Shuttle Radar Topography Mission (SRTM). This means one can view almost the entire earth in three dimensions. Since November 2006, the 3D views of many mountains, including Mount Everest, have been improved by the use of supplementary DEM data to fill the gaps in SRTM coverage.

Some people use the applications to add their own data, making them available through various sources, such as the Bulletin Board Systems (BBS) or blogs mentioned in the link section below. Google Earth is able to show various kinds of images overlaid on the surface of the earth and is also a Web Map Service client. Google Earth supports managing three-dimensional Geospatial data through Keyhole Markup Language (KML).

Detail

Google Earth is simply based on 3D maps, with the capability to show 3D buildings and structures (such as bridges), which consist of users' submissions using SketchUp, a 3D modeling program software. In prior versions of Google Earth (before Version 4), 3D buildings were limited to a few cities, and had poorer rendering with no textures. Many buildings and structures from around the world now have detailed 3D structures; including (but not limited to) those in the United States, Canada, Mexico, India, Japan, United Kingdom, Spain, Germany, Pakistan and the cities, Amsterdam and Alexandria. In August 2007, Hamburg became the first city entirely shown in 3D, including textures such as façades. The 'Westport3D' model was created by 3D imaging firm AM3TD using long-distance laser scanning technology and digital photography and is the first such model of an Irish town to be created. As it was developed initially to aid Local Government in carrying out their town planning functions it includes the highest-resolution photo-realistic textures to be found anywhere in Google Earth. Three-dimensional renderings are available for certain buildings and structures around the world via Google's 3D Warehouse and other websites. In June 2012, Google announced that it will start to replace user submitted 3D buildings with auto-generated 3D mesh buildings starting with major cities. Although there are many cities on Google Earth that are fully or partially 3D, more are available in the Earth Gallery. The Earth Gallery is a library of modifications of Google Earth people have made. In the library there are not only modifications for 3D buildings, but also models of earthquakes using the Google Earth model, 3D forests, and much more.

Recently, Google added a feature that allows users to monitor traffic speeds at loops located every 200 yards in real-time. In 2007, Google began offering traffic data in real-time, based on information crowdsourced from the GPS-identified locations of cellular phone users. In version 4.3 released on April 15, 2008, Google Street View was fully integrated into the program allowing the program to provide an on the street level view in many locations.

On January 31, 2010, the entirety of Google Earth's ocean floor imagery was updated to new images by SIO, NOAA, US Navy, NGA, and GEBCO. The new images have caused smaller islands, such as some atolls in the Maldives, to be rendered invisible despite their shores being completely outlined.

Uses

Google Earth may be used to perform some day-to-day tasks and for other purposes.

- Google Earth can be used to view areas subjected to widespread disasters if Google supplies up-to-date images. For example, after the January 12, 2010 Haiti earthquake images of Haiti were made available on January 17.

- With Google's push for the inclusion of Google Earth in the Classroom, teachers are adopting Google Earth in the classroom for lesson planning, such as teaching students geographical themes (location, culture, characteristics, human interaction, and movement) to creating mashups with other web applications such as Wikipedia.

- One can explore and place location bookmarks on the Moon and Mars.

- One can get directions using Google Earth, using variables such as street names, cities, and establishments. But the addresses must by typed in search field, one cannot simply click on two spots on the map.

- Google Earth can function as a hub of knowledge, pertaining the users location. By enabling certain options, one can see the location of gas stations, restaurants, museums, and other public establishments in their area. Google Earth can also dot the map with links to images, YouTube videos, and Wikipedia articles relevant to the area being viewed.

- One can create custom image overlays for planning trips, hikes on handheld GPS units.

- Google Earth can be used to map homes and select a random sample for research in developing countries.

All of these features are also released by Google Earth Blog.

Features

Wikipedia and Panoramio Integration

In December 2006, Google Earth added a new layer called "Geographic Web" that includes integration with Wikipedia and Panoramio. In Wikipedia, entries are scraped for coordinates via the Coord templates. There is also a community-layer from the project Wikipedia-World. More coordinates are used, different types are in the display and different languages are supported than the built-in Wikipedia layer. Google announced on May 30, 2007 that it is acquiring Panoramio. In March 2010, Google removed the "Geographic Web" layer. The "Panoramio" layer became part of the main layers and the "Wikipedia" layer was placed in the "More" layer.

Flight Simulator

Downtown Toronto as seen from a F-16 Fighting Falcon during a simulated flight.

In Google Earth v4.2 a flight simulator was included as a hidden feature. Starting with v4.3 it is no longer hidden. The flight simulator could be accessed by holding down the keys Ctrl, Alt, and A. Initially the F-16 Fighting Falcon and the Cirrus SR-22 were the only aircraft available, and they could be used with only a few airports. However, one can start flight in "current location" and need not to be at an airport. One will face the direction they face when they start the flight simulator. They cannot start flight in ground level view and must be near the ground (approximately 50m-100m above the ground) to start in take-off position. Otherwise they will be in the air with 40% flaps and gears extended (landing position). In addition to keyboard control, the simulator can be controlled with a mouse or joystick. Google Earth v5.1 and higher crashes when starting flight simulator with Saitek and other joysticks. The user can also fly underwater.

Featured Planes

- F-16 Fighting Falcon – A much higher speed and maximum altitude than the Cirrus SR-22, it has the ability to fly at a maximum speed of Mach 2, although a maximum speed of 1678 knots (3108 km/h) can be achieved. The take-off speed is 225 knots, the landing speed is 200 knots (370 km/h).

- Cirrus SR-22 – Although slower and with a lower maximum altitude, the SR-22 is much easier to handle and is preferred for up-close viewing of Google Earth's imagery. The take-off speed is 75 knots (139 km/h), the landing speed is 70 knots (130 km/h).

The flight simulator can be commanded with the keyboard, mouse or plugged-in joystick. Broadband connection and a high speed computer provides a very realistic experience. The simulator also runs with animation, allowing objects (for example: planes) to animate while on the simulator. Programming language can also be used to make it look like the cockpit of a plane, or for instrument landing.

Sky Mode

Google Sky is a feature that was introduced in Google Earth 4.2 on August 22, 2007, and allows users to view stars and other celestial bodies. It was produced by Google through a partnership with the Space Telescope Science Institute (STScI) in Baltimore, the science operations center for the Hubble Space Telescope. Dr. Alberto Conti and his co-developer Dr. Carol Christian of STScI plan to add the public images from 2007, as well as color images of all of the archived data from Hubble's Advanced Camera for Surveys. Newly released Hubble pictures will be added to the Google Sky program as soon as they are issued. New features such as multi-wavelength data, positions of major satellites and their orbits as well as educational resources will be provided to the Google Earth community and also through Christian and Conti's website for Sky. Also visible on Sky mode are constellations, stars, galaxies and animations depicting the planets in their orbits. A real-time Google Sky mashup of recent astronomical transients, using the VOEvent protocol, is being provided by the VOEventNet collaboration. Google's Earth maps are being updated each 5 minutes.

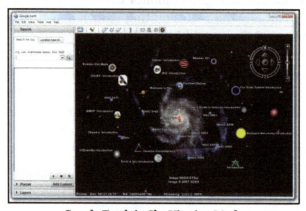
Google Earth in Sky Viewing Mode.

Google Sky faces competition from Microsoft WorldWide Telescope (which runs only under the Microsoft Windows operating systems) and from Stellarium, a free open source planetarium that runs under Microsoft Windows, OS X, and Linux.

On March 13, 2008, Google made a web-based version of Google Sky available via the internet.

Street View

On April 15, 2008 with version 4.3, Google fully integrated its Street View into Google Earth. In version 6.0, the photo zooming function has been removed because it is incompatible with the new 'seamless' navigation.

Google Street View provides 360° panoramic street-level views and allows users to view parts of selected cities and their surrounding metropolitan areas at ground level. When it was launched on May 25, 2007 for Google Maps, only five cities were included. It has since expanded to more than 40 U.S. cities, and includes the suburbs of many, and in some cases, other nearby cities. Recent updates have now implemented Street View in most of the major cities of Canada, Mexico, Denmark, South Africa, Japan, Spain, Norway, Finland, Sweden, France, the UK, Republic of Ireland, the Netherlands, Italy, Switzerland, Portugal, Taiwan, and Singapore.

Google Street View, when operated, displays photos that were previously taken by a camera mounted on an automobile, and can be navigated by using the mouse to click on photograph icons displayed on the screen in the user's direction of travel. Using these devices, the photos can be viewed in different sizes, from any direction, and from a variety of angles.

Water and Ocean

Introduced in version 5.0 (February 2009), the *Google Ocean* feature allows users to zoom below the surface of the ocean and view the 3D bathymetry beneath the waves. Supporting over 20 content layers, it contains information from leading scientists and oceanographers. On April 14, 2009, Google added underwater terrain data for the Great Lakes. In 2010, Google added underwater terrain data for Lake Baikal.

In June 2011, higher resolution of some deep ocean floor areas increased in focus from 1-kilometer grids to 100 meters thanks to a new synthesis of seafloor topography released through Google Earth. The high-resolution features were developed by oceanographers at Columbia University's Lamont-Doherty Earth Observatory from scientific data collected on research cruises. The sharper focus is available for about 5 percent of the oceans (an area larger than North America). Underwater scenery can be seen of the Hudson Canyon off New York City, the Wini Seamount near Hawaii, and the sharp-edged 10,000-foot-high Mendocino Ridge off the U.S Pacific Coast. There is a Google 2011 Seafloor Tour for those interested in viewing ocean deep terrain.

Historical Imagery

Introduced in version 5.0, Historical Imagery allows users to traverse back in time and study earlier stages of any place. This feature allows research that require analysis of past records of various places.

Mars

A picture of Mars' landscape.

Google Earth 5 includes a separate globe of the planet Mars, that can be viewed and analysed for research purposes. The maps are of a much higher resolution than those on the browser version of Google Mars and it also includes 3D renderings of the Martian terrain. There are also some extremely-high-resolution images from the Mars Reconnaissance Orbiter's HiRISE camera that are of a similar resolution to those of the cities on Earth. Finally, there are many high-resolution panoramic images from various Mars landers, such as the Mars Exploration Rovers, Spirit and Opportunity, that can be viewed in a similar way to Google Street View. Interestingly enough, layers on Google Earth (such as World Population Density) can also be applied to Mars. Layers of Mars can also be applied onto Earth. Mars also has a small application found near the face on Mars. It is called Meliza, and features a chat between the user and an automatic robot speaker. It is useful for research on Mars, but is not recommended for normal conversations.

Moon

One of the lunar landers viewed in Google Moon.

On July 20, 2009, the 40th anniversary of the Apollo 11 mission, Google introduced the Google Earth version of Google Moon, which allows users to view satellite images of the Moon. It was announced and demonstrated to a group of invited guests by Google along with Buzz Aldrin at the Newseum in Washington, D.C.

Google Earth Engine

Google Earth Engine is a separate product, not a part of Google Earth.

Liquid Galaxy

Liquid Galaxy is a cluster of computers running Google Earth creating an immersive experience. On September 30, 2010, Google made the configuration and schematics for their rigs public, placing code and setup guides on the Liquid Galaxy wiki.

Liquid Galaxy has also been used as a panoramic photo viewer using KRpano, as well as a Google Street View viewer using Peruse-a-Rue Peruse-a-Rue is a method for synchronizing multiple Maps API clients.

Influences

Google Earth can be traced directly back to a small company named Autometric, now a part of Boeing. A team at Autometric, led by Robert Cowling, created a visualization product named Edge Whole Earth. Bob demonstrated Edge to Michael T. Jones, Chris Tanner and others at SGC in 1996. Several other visualization products using imagery existed at the time, including Performer-based ones, but Michael T. Jones stated emphatically that he had "never thought of the complexities of rendering an entire globe …" The catch phrase "from outer space to in your face" was coined by Autometric President Dan Gordon, and used to explain his concept for personal/local/global range. Edge blazed a trail as well in broadcasting, being used in 1997 on CBS News with Dan Rather, in print for rendering large images draped over terrain for National Geographic, and used for special effects in the feature film *Shadow Conspiracy* in 1997.

Gordon was a huge fan of the 'Earth' program described in Neal Stephenson's sci-fi classic *Snow Crash*. Indeed, a Google Earth co-founder claimed that Google Earth was modeled after *Snow Crash*, while another co-founder said it was inspired by the short science education film *Powers of Ten*. In fact Google Earth was at least partly inspired by a Silicon Graphics demo called "From Outer Space to in Your Face" which zoomed from space into the Swiss Alps then into the Matterhorn. This launch demo was hosted by an Onyx 3000 with InfiniteReality4 graphics, which supported Clip Mapping and was inspired by the hardware texture paging capability (although it did not use the Clip Mapping) and "Powers of Ten". The first Google Earth implementation called Earth Viewer emerged from Intrinsic Graphics as a demonstration of Chris Tanner's software based implementation of a Clip Mapping texture paging system and was spun off as Keyhole Inc.

Versions and Variations

Mac Version

Since version 4.1.7076.4558 (released on May 9, 2007) onward OS X users can, among other new features, upgrade to the "Plus" version via an option in the Google Earth menu. Some users reported difficulties with Google Earth crashing in the then current version when zooming in. Version 5 of Google Earth for Mac was released in 2009, and version 7 was released concomitantly with the Mac and PC versions on 31 October 2012.

Linux Version

Starting with the version 4 beta Google Earth functions under Linux, as a native port using the Qt toolkit. The Free Software Foundation consider the development of a free compatible client for Google Earth to be a High Priority Free Software Project.

Android Version

Google Earth running on Android.

An Android version was released on Monday, February 22, 2010.

iOS Version

A version for the iOS, which runs on the iPhone, iPod Touch and the iPad, was released for free on the App Store on October 27, 2008. It makes use of the multi-touch interface to move on the globe, zoom or rotate the view, and allow to select the current location using the iPhone integrated Assisted GPS. Although it previously did not support any layers apart from Wikipedia and Panoramio, version 6.2 brought KML support to add additional layers. Version 7 introduced 3D modeling of several cities.

Google Earth Plus (Discontinued in 2008)

Discontinued in December 2008, Google Earth Plus was an individual-oriented paid subscription upgrade to Google Earth that provided customers with the following features, most of which are now available in the free Google Earth.

- GPS integration: read tracks and waypoints from a GPS device. A variety of third-party applications have been created which provide this functionality using the basic version of Google Earth by generating KML or KMZ files based on user-specified or user-recorded waypoints. However, Google Earth Plus provides direct support for the Magellan and Garmin product lines, which together hold a large share of the GPS market. The Linux version of the Google Earth Plus application does not include any GPS functionality.

- Higher-resolution printing.
- Customer support via email.
- Data importer: read address points from CSV files; limited to 100 points/addresses. A feature allowing path and polygon annotations, which can be exported to KML, was formerly only available to Plus users, but was made free in version 4.0.2416.
- Higher data download speeds

Google Earth Pro

Google Earth Pro is a business-oriented upgrade to Google Earth that has more features than the Plus version. It is the most feature-rich version of Google Earth available to the public, with various additional features such as a movie maker and data importer. In addition to business-friendly features, it has also been found useful for travelers with map-making tools. Up until late January 2015, it was available for $399/year, however Google decided to make it free to the public. It is now for free and Google does not mention anything about new policy changes. The Pro version includes add-on software such as:

- Movie making.
- GIS data importer.
- Advanced printing modules.
- Radius and area measurements.

Google Earth Pro is available for Windows (NT-based versions), Mac OS X 10.4 or later.

Google Earth Enterprise

Google Earth Enterprise is a version of Google Earth designed for use by organizations whose businesses could take advantage of the program's capabilities, for example by having a globe that holds company data available for anyone in that company. As of March 20, 2015 Google has retired the Google Earth Enterprise product, with support ending March 22, 2017.

Automotive Version

An automotive version of Google Earth is available in the 2010 Audi A8.

Google Earth Plug-in

The Google Earth API is a free beta service, available for any web site that is free to consumers. The Plug-in and its JavaScript API let users place a version of Google Earth into web pages. The API enables sophisticated 3D map applications to be built. At its unveiling at Google's 2008 I/O developer conference, the company showcased potential applications such as a game where the player controlled a milktruck atop a Google Earth surface.

The Google Earth API has been deprecated as of 15 December 2014 and will remain supported until the 15th of December 2015. Google Chrome aims to end support for the Netscape Plugin API (which the Google Earth API relies on) by the end of 2016.

Controversy and Criticism

The software has been criticized by a number of special interest groups, including national officials, as being an invasion of privacy and even posing a threat to national security. The typical argument is that the software provides information about military or other critical installations that could be used by terrorists.

- Former President of India APJ Abdul Kalam expressed concern over the availability of high-resolution pictures of sensitive locations in India. Google subsequently agreed to censor such sites.

- The Indian Space Research Organisation said Google Earth poses a security threat to India, and seeks dialogue with Google officials.

- The South Korean government expressed concern that the software offers images of the presidential palace and various military installations that could possibly be used by hostile neighbor North Korea.

- In 2006, one user spotted a large topographical replica in a remote region of China. The model is a small-scale (1/500) version of the Karakoram Mountain Range, which is under the control of China but claimed by India. When later confirmed as a replica of this region, spectators began entertaining military implications.

- In 2006, Google Earth began offering detailed images of classified areas in Israel. The images showed Israel Defense Forces bases, including secret Israeli Air Force facilities, Israel's Arrow missile defense system, military headquarters and Defense Ministry compound in Tel Aviv, a top-secret power station near Ashkelon, and the Negev Nuclear Research Center. Also shown was the alleged headquarters of the Mossad, Israel's foreign intelligence service, whose location is highly classified.

- Operators of the Lucas Heights nuclear reactor in Sydney, New South Wales,

Australia asked Google to censor high-resolution pictures of the facility. However, they later withdrew the request.

- In July 2007, it was reported that a new Chinese Navy Jin-class nuclear ballistic missile submarine was photographed at the Xiaopingdao Submarine Base south of Dalian.

- Hamas and the al-Aqsa Martyrs' Brigades have reportedly used Google Earth to plan Qassam rocket attacks on Israel from Gaza.

- The lone surviving gunman involved in the 2008 Mumbai attacks admitted to using Google Earth to familiarise himself with the locations of buildings used in the attacks.

- Michael Finton, aka Talib Islam, used Google Earth in planning his attempted September 24, 2009, bombing of the Paul Findley Federal Building and the adjacent offices of Congressman Aaron Schock in Springfield, Illinois.

- In 2009, Google superimposed old woodblock prints of maps from 18th and 19th century Japan over Japan today. These maps marked areas inhabited by the burakumin caste, who were considered "non-humans" for their "dirty" occupations, including leather tanning and butchery. Descendants of members of the burakumin caste still face discrimination today and many Japanese people feared that some would use these areas, labeled *etamura* (穢多村, *translation: "village of an abundance of defilement"*), to target current inhabitants of them. These maps are still visible on Google Earth, but with the label removed where necessary.

Countries where Google Earth is Blocked:				
Country	**By Whom**	**Reason**	**Since When**	**Source**
Iran	Google	US government export restrictions	2007	
Morocco	Maroc Telecom, the most popular service provider	Unknown	2006	
Sudan	Google	US government export restrictions	2007	

Google Earth has been blocked by Google in Iran and Sudan since 2007 due to US government export restrictions. The program has also been blocked in Morocco since 2006 by Maroc Telecom, a major service provider in the country.

Some citizens may express concerns over aerial information depicting their properties and residences being disseminated freely. As relatively few jurisdictions actually guarantee the individual's right to privacy, as opposed to the state's right to secrecy, this is an evolving point. Perhaps aware of these critiques, for a time, Google had Area 51 (which is highly visible and easy to find) in Nevada as a default placemark when Google Earth is first installed.

Blurred out image of the Royal Stables in The Hague, Netherlands. This has since been partially lifted.

As a result of pressure from the United States government, the residence of the Vice President at Number One Observatory Circle was obscured through pixelization in Google Earth and Google Maps in 2006, but this restriction has since been lifted. The usefulness of this downgrade is questionable, as high-resolution photos and aerial surveys of the property are readily available on the Internet elsewhere. Capitol Hill also used to be pixelized in this way. The Royal Stables in The Hague, Netherlands also used to be pixelized, and are still pixelized at high zoom levels.

Critics have expressed concern over the willingness of Google to cripple their dataset to cater to special interests, believing that intentionally obscuring any land goes against its stated goal of letting the user "point and zoom to any place on the planet that you want to explore".

In the United Kingdom, critics have also argued that Google Earth has led to the vandalism of private property, highlighting the graffiti of a penis being drawn on the roof of a house near Hungerford, on the roof of Yarm School at Stockton on Tees and on the playing fields of a school in Southampton as examples of this.

In Hazleton, Pennsylvania, media attention and critics focused on Google Earth once more because of the defacing of the Hazleton Area High School football field. Grass was removed to create the image of a penis approximately 35 yards long and 20 yards wide.

Late 2000s versions of Google Earth require a software component running in the background that will automatically download and install updates. Several users expressed concerns that there is not an easy way to disable this updater, as it runs without the permission of the user.

In the academic realm increasing attention has been devoted to both Google Earth and its place in the development of digital globes more generally. In particular, the International Journal of Digital Earth now features many articles evaluating and comparing the development Google Earth and its differences when compared to other professional, scientific and governmental platforms.

Elsewhere, in the Humanities and Social Sciences, Google Earth's role in the expansion of "earth observing media" has been examined. Leon Gurevitch in particular has examined the role of Google Earth in shaping a shared cultural consciousness regarding climate change and humanity's capacity to treat the earth as an engineerable object. Gurevitch has described this interface between earth representation in Google Earth and a shared cultural imaginary of geo-engineering as "Google Warming".

Copyright

Every image created from Google Earth using satellite data provided by Google Earth is a copyrighted map. Any derivative from Google Earth is made from copyrighted data which, under United States Copyright Law, may not be used except under the licenses Google provides. Google allows non-commercial personal use of the images (e.g. on a personal website or blog) as long as copyrights and attributions are preserved. By contrast, images created with NASA's globe software World Wind use The Blue Marble, Landsat or USGS layer, each of which is a terrain layer in the public domain. Works created by an agency of the United States government are public domain at the moment of creation. This means that those images can be freely modified, redistributed and used for commercial purposes.

Layers

Google Earth also features many layers as a source for information on businesses and points of interest, as well as showcasing the contents of many communities, such as Wikipedia, Panoramio and YouTube.

Borders and Labels

Contains borders for countries/provinces and shows placemarks for cities and towns.

- Borders: *Marks international borders with a thick yellow line (borders with territorial disputes with thick red lines), 1st level administrative borders (generally provinces and states) with a lavender line, and 2nd level administrative borders (counties) with a cyan line. Coastlines appear as a thin yellow line. Displays names of countries, 1st level administrative areas, and islands.*
- Labels: *Displays labels for large bodies of water, such as oceans, seas, and bays, and populated places.*

3D Imagery

Google Earth 3D shows many 3D computer graphics building models in many cities, in these styles:

- 3D trees: *Shows many trees in Athens, Greece; Surui Forest, Brazil; Kahigaini,*

Kenya; Mangrove Forests, Mexico; Jedediah Smith Redwoods State Park, California.

- Photorealistic: *Shows many buildings in a realistic style, with more complex polygons and surface images.*

- Autogen: *Renders entire metropolitan areas in 3D via processing of 45 degree aerial imagery.*

- Gray: *Low-detail models of city buildings designed for computers that may not have the capability of showing the photorealistic models.*

In 2009, in a unique collaboration between Google and the Museo del Prado in Madrid, the museum selected 14 of its most important paintings to be photographed and displayed at the ultrahigh resolution of 14,000 megapixels inside the 3D version of the Prado in Google Earth and Google Maps.

3D coverage in Google Earth.

In June 2012, Google announced that it will be replacing user made 3D buildings with an auto-generated 3D mesh. This will be phased in, starting with select larger cities, with the notable exception of cities such as London and Toronto which require more time to process detailed imagery of their vast number of buildings. The reason given is to have greater uniformity in 3D buildings, and to compete with other platforms already using the technology such as Nokia Here and Apple Maps.

The first 3D buildings in Google Earth were created using 3D modeling applications such as SketchUp and, beginning in 2009, Building Maker, and were uploaded to Google Earth via the 3D Warehouse.

In 2012, Google began incorporating automatically-generated 3D imagery, which displays entire areas in 3D rather than individual buildings, into the mobile and desktop versions of Google Earth, releasing coverage of 21 cities in four countries that year. By March 2015, 3D imagery covering more than 300,000 km² was available and by early

2016 had been expanded to hundreds of cities in over 40 countries, including every U.S. state and encompassing every continent except Antarctica.

During 2015, Hong Kong and places in the Philippines were added to the coverage.

As of February 2016, 3D imagery covering more than 495,000 km² was available in cities in over 40 countries, covering all continents except Antarctica.

Google Street View

Shows placemarks with 360 degree panoramic views of streets of many cities in Australia, France, the United Kingdom, Republic of Ireland, Italy, Japan, New Zealand, Spain, the United States, and recently Portugal, Brazil, the Netherlands, Taiwan, Switzerland, Canada, Mexico, Sweden, Norway, South Africa and Finland.

Weather

- Clouds – *Displays cloud cover based on data from both geostationary and low Earth-orbiting satellites. The clouds appear at their calculated elevation, determined by measuring the cloud top temperature relative to surface temperature.*

- Radar – *Displays weather radar data provided by weather.com and Weather Services International, updating every 5–6 minutes.*

- Conditions and Forecast – *Displays local temperatures and weather conditions. Clicking on an indicator displays a 2 Day Forecast (Example: Monday Morning, Monday Night, Tuesday Morning, Tuesday Night) forecast provided by weather.com.*

- Information – *Clicking Information allows users to further read up on where Google Earth gets weather information.*

Sky Layers

Layers for Google Sky:

- EarthSky Podcasts.

References

- O'Leary, Beth Laura; Darrin, Ann Garrison (2009). Handbook of Space Engineering, Archaeology, and Heritage. Hoboken: CRC Press. pp. 239–240. ISBN 9781420084320
- Haque, Akhlaque (2015). Surveillance, Transparency and Democracy: Public Administration in the Information Age. Tuscaloosa, AL: University of Alabama Press. pp. 70–73. ISBN 978-0817318772

- McNamara, Joel (2008). GPS For Dummies. John Wiley & Sons. p. 59. ISBN 0-470-45785-6., // books.google.com/books?id=Hbz4LYIrvuMC&pg=PA59

- Misra, Pratap; Enge, Per (2006). Global Positioning System. Signals, Measurements and Performance (2nd ed.). Ganga-Jamuna Press. p. 115. ISBN 0-9709544-1-7. Retrieved 19, January 2020

- Samama, Nel (2008). Global Positioning: Technologies and Performance. John Wiley & Sons. p. 65. ISBN 0-470-24190-X.,

- Michael Russell Rip; James M. Hasik (2002). The Precision Revolution: GPS and the Future of Aerial Warfare. Naval Institute Press. ISBN 1-55750-973-5. Retrieved 22, August 2020

- Dietrich Schroeer; Mirco Elena (2000). Technology Transfer. Ashgate. p. 80. ISBN 0-7546-2045-X. Retrieved 15, June 2020

- Michael Russell Rip; James M. Hasik (2002). The Precision Revolution: GPS and the Future of Aerial Warfare. Naval Institute Press. p. 65. ISBN 1-55750-973-5. Retrieved 16, February 2020

Chapter 5
Uses of Remote Sensing

Remote sensing has emerged as an effective tool for analysis and better management of natural resources. It has applications in various fields such as irrigation management, flood mapping, environmental monitoring, runoff model etc. The diverse applications of remote sensing in the current scenario have been thoroughly discussed in this chapter.

Watershed

Scientific planning and management is essential for the conservation of land and water resources for optimum productivity. Watersheds being the natural hydrologic units, such studies are generally carried out at watershed scale and are broadly referred under the term watershed management. It involves assessment of current resources status, complex modeling to assess the relationship between various hydrologic components, planning and implementation of land and water conservation measures etc.

Remote sensing via aerial and space-borne platforms acts as a potential tool to supply the essential inputs to the land and water resources analysis at different stages in watershed planning and management. Water resource mapping, land cover classification, estimation of water yield and soil erosion, estimation of physiographic parameters for land prioritization and water harvesting are a few areas where remote sensing techniques have been used.

Various remote sensing applications in water resources management under the following five classes:

- Water resources mapping.
- Estimation of watershed physiographic parameters.
- Estimation of hydrological and meteorological variables.
- Watershed prioritization.
- Water conservation.

Water Resources Mapping

Identification and mapping of the surface water boundaries has been one of the simplest and direct applications of remote sensing in water resources studies. Water resources mapping using remote sensing data require fine spatial resolution so as to achieve accurate delineation of the boundaries of the water bodies.

Optical remote sensing techniques, with their capability to provide very fine spatial resolution have been widely used for water resources mapping. Water absorbs most of the energy in NIR and MIR wavelengths giving darker tones in the bands, and can be easily differentiated from the land and vegetation.

The following figure shows images of a part of the Krishna river basin in different bands of the Landsat ETM+. In the VIS bands (bands 1, 2 and 3) the contrast between water and other features are not very significant. On the other hand, the IR bands (bands 4 and 5) show a sharp contrast between them due to the poor reflectance of water in the IR region of the EMR spectrum.

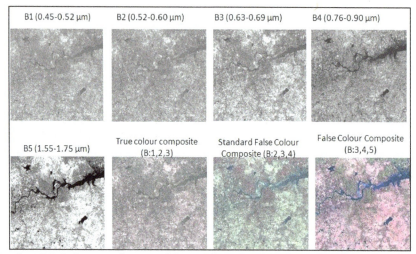

Landsat ETM+ images of a part of the Krishna river basin in different spectral bands (Nagesh Kumar and Reshmidevi, 2013).

Poor cloud penetration capacity and poor capability to map water resources under thick vegetation cover are the major drawbacks of the optical remote sensing techniques.

Use of active microwave sensor helps to overcome these limitations as the radar waves can penetrate the clouds and the vegetation cover to some extent. In microwave remote sensing, water surface provides specular reflection of the microwave radiation, and hence very little energy is scattered back compared to the other land features. The difference in the energy received back at the radar sensor is used for differentiating, and to mark the boundaries of the water bodies.

Estimation of Watershed Physiographic Parameters

This section covers the remote sensing applications in estimating watershed physiographic parameters and the land use / land cover information.

Watershed Physiographic Parameters

Various watershed physiographic parameters that can be obtained from remotely sensed data include watershed area, size and shape, topography, drainage pattern and landforms.

Stereoscopic attribute of aerial photographs or satellite images permit quantitative assessment of landforms and evaluation of basin topography, which can be used to develop or update the topographic maps. With the help of satellite remote sensing, global scale digital elevation models (DEMs) are available today at fine spatial resolution and reasonable vertical accuracy. DEM from the Shuttle Radar Topographic Mission (SRTM) and ASTER GDEM are examples. SRTM DEM provides near-global DEM at 90m spatial resolution and 16m vertical accuracy. Airborne laser altimeters also provide quick and accurate measurements for evaluating changes in land surface features and are effective tools to ascertain watershed properties.

Fine resolution DEMs have been used to extract the drainage network/ pattern using the flow tracing algorithms. The drainage information can also be extracted from the optical images using digital image processing techniques.

The drainage information may be further used to generate secondary information such as structure of the basin, basin boundary, stream orders, stream length, stream frequency, bifurcation ratio, stream sinuosity, drainage density and linear aspects of channel systems etc.

The figure below shows the ASTER GDEM for a small region in the Krishna Basin in North Karnataka and the drainage network delineated from it using the flow tracing algorithm included in the 'spatial analyst' tool box of ArcGIS. Fig.(b) also shows the stream orders assigned to each of the delineated streams.

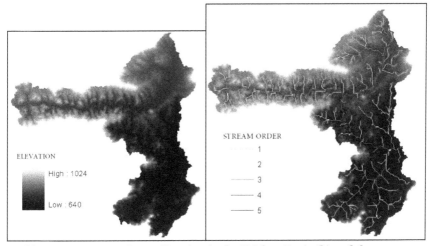

(a) ASTER GDEM of a small region in the Krishna Basin (b) and the stream network delineated from the DEM.

Land use / Land Cover Classification

Detailed land use / land cover map is another important input that remote sensing can yield for hydrologic analysis.

Land cover classification using multispectral remote sensing data is one of the earliest, and well established remote sensing applications in water resources studies. With the capability of the remote sensing systems to provide frequent temporal sampling and the fine spatial resolution, it is possible to analyze the dynamics of land use / land cover pattern, and also its impact on the hydrologic processes.

Use of hyper-spectral imageries helps to achieve further improvement in the land use / land cover classification, wherein the spectral reflectance values recorded in the narrow contiguous bands are used to differentiate different land use classes which show close resemblance with each other. Identification of crop types using hyper-spectral data is an example.

With the help of satellite remote sensing, land use land cover maps at near global scale are available today for hydrological applications. European Space Agency (ESA) has released a global land cover map of 300 m resolution, with 22 land cover classes at 73% accuracy.

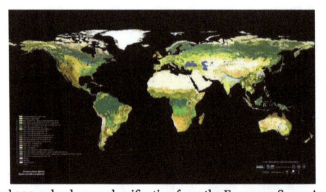

Global 300 m land cover classification from the European Space Agency.

Estimation of Hydrological and Meteorological Variables

Hydrological processes such as precipitation and evapotranspiration are generally used as inputs to the hydrological models to simulate other processes such as runoff (surface and sub- surface), storage change in the unsaturated zone, and ground water flow. This section covers the remote sensing applications in estimating precipitation, evapotranspiration and soil moisture.

Precipitation

Remote sensing techniques have been used to provide information about the occurrence of rainfall and its intensity. Basic concept behind the satellite rainfall estimation is the differentiation of precipitating clouds from the non-precipitating clouds (Gibson and Power, 2000) by relating the brightness of the cloud observed in the imagery to the rainfall intensities.

Satellite remote sensing uses both optical and microwave remote sensing (both passive and active) techniques.

The following table lists some of the important satellite rainfall data sets, satellites used for the data collection and the organizations that control the generation and distribution of the data.

Table: Details of some of the important satellite rainfall products (Nagesh Kumar and Reshmidevi, 2013).

Program	Organization	Spectral bands used	Characteristics and source of data
World Weather Watch	WMO	VIS, IR	1-4 km spatial, and 30 min. temporal resolution
TRMM	NASA JAXA	VIS, IR Passive & active microwave	Sub-daily 0.25° (~27 km) spatial resolution
PERSIANN	CHRS	IR	0.25° spatial resolution Temporal resolution: 30 min. aggregated to 6 hrs.
CMORPH	NOAA	Microwave	0.08 deg (8 km) spatial and 30 min. temporal resolution
Acronyms			
CHRS: Center for Hydrometeorology and Remote Sensing, University of California, USA CMORPH: Climate Prediction Center (CPC) MORPHing technique.			
NASA: National Aeronautics and Space Administration, USA NOAA: National Oceanic and Atmospheric Administration, USA.			
PERSIANN: Precipitation Estimation from Remotely Sensed Information using Artificial Neural Network TRMM: Tropical Rainfall Measuring Mission.			
WMO: World Meteorological Organization.			

Evapotranspiration

Evapotranspiration (ET) represents the water and energy flux between the land surface and the lower atmosphere. ET fluxes are controlled by the feedback mechanism between the atmosphere and the land surface, soil and vegetation characteristics, and the hydro- meteorological conditions.

There are no direct methods available to estimate the actual ET by means of remote sensing techniques. Remote sensing application in the ET estimation is limited to the estimation of the surface conditions like albedo, soil moisture, surface temperature, and vegetation characteristics like normalized differential vegetation index (NDVI) and leaf area index (LAI). The data obtained from remote sensing are used in different models to simulate the actual ET.

Courault et al. (2005) grouped the remote sensing data-based ET models into four different classes:

- Empirical direct methods: Use the empirical equations to relate the difference in the surface air temperature to the ET.

- Residual methods of the energy budget: Use both empirical and physical parameterization. Example: SEBAL (Bastiaanssen et al., 1998), FAO-56 method (Allen at al., 1998).

- Deterministic models: Simulate the physical process between the soil, vegetation and atmosphere making use of remote sensing data such as Leaf Area Index (LAI) and soil moisture. SVAT (Soil-Vegetation-Atmosphere-Transfer) model is an example (Olioso et al., 1999).

- Vegetation index methods: Use the ground observation of the potential or reference ET. Actual ET is estimated from the reference ET by using the crop coefficients obtained from the remote sensing data (Allen et al., 2005; Neale et al., 2005).

Optical remote sensing using the VIS and NIR bands have been commonly used to estimate the input data required for the ET estimation algorithms.

As a part of the NASA / EOS project to estimate global terrestrial ET from earth's land surface by using satellite remote sensing data, MODIS Global Terrestrial Evapotranspiration Project (MOD16) provides global ET data sets at regular grids of 1 sq.km for the land surface at 8-day, monthly and annual intervals for the period 2000-2010.

Soil Moisture Estimation

Remote sensing techniques of soil moisture estimation are advantageous over the conventional *in-situ* measurement approaches owing to the capability of the sensors to capture spatial variation over a large aerial extent. Moreover, depending upon the revisit time of the satellites, frequent sampling of an area and hence more frequent soil moisture measurements are feasible.

The following figure shows the global average monthly soil moisture in May extracted from the integrated soil moisture database of the European Space Agency- Climate Change Initiative (ESA-CCI).

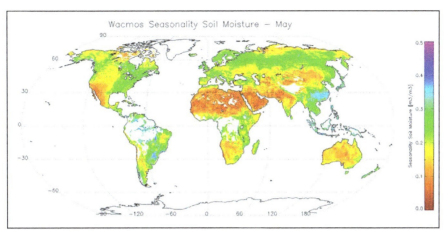
Global monthly average soil moisture in May from the CCI data.

Remote sensing of the soil moisture requires information below the ground surface and therefore mostly confined to the use of thermal and microwave bands of the EMR spectrum.

Remote sensing of the soil moisture is based on the variation in the soil properties caused due to the presence of water. Soil properties generally monitored for soil moisture estimation include soil dielectric constant, brightness temperature, and thermal inertia.

Though the remote sensing techniques are giving reasonably good estimation of the soil moisture, due to the poor surface penetration capacity of the microwave signals, it is considered to be effective in retrieving the moisture content of the surface soil layer of maximum 10 cm thickness. In the recent years, attempts have been made to extract the soil moisture of the entire root zone with the help of remote sensing data. Such methods assimilate the remote sensing derived surface soil moisture data with physically based distributed models to simulate the root zone soil moisture. For example, Das et al. (2008) used the Soil-Water-Atmosphere-Plant (SWAP) model for simulating the root zone soil moisture by assimilating the aircraft-based remotely sensed soil moisture into the model.

Some of the satellite based sensors that have been used for retrieving the soil moisture information are the following:

- Passive microwave sensors: SMMR, AMSR-E and SSM/I.
- Active microwave sensors (radar): Advanced SCATterometer (ASCAT) aboard the EUMETSAT MetOp satellite.
- Thermal sensors: Data from the thermal bands of the MODIS sensor onboard Terra satellite have also been used for retrieving soil moisture data.

Use of hyper-spectral remote sensing technique has been recently employed to improve the soil moisture simulation. Hyper-spectral monitoring of the soil moisture uses reflectivity in the VIS and the NIR bands to identify the changes in the spectral reflectance curves due to the presence of soil moisture (Yanmin et al., 2010). Spectral reflectance measured in multiple narrow bands in the hyperspectral image helps to extract most appropriate bands for the soil moisture estimation, and to identify the changes in the spectral reflectance curves due to the presence of soil moisture.

Watershed characterization and Prioritization

Watershed characterization involves the measurement and analysis of various hydro-geological and geo-morphological parameters, soil and land use characteristics etc. (Rao and Raju, 2010).

Watershed prioritization is the ranking of different watersheds or sub-watersheds within a watershed for any specific application based on the watershed characteristics.

Water Conservation and Rainwater Harvesting

Rainwater harvesting, wherein water from the rainfall is stored for future usage, is an effective water conservation measure particularly in the arid and semi-arid regions.

Rainwater harvesting techniques are highly location specific. Selection of appropriate water harvesting technique requires extensive field analysis to identify the rainwater harvesting potential of the area, and the physiographic and terrain characteristics of the locations. It depends on the amount of rainfall and its distribution, land topography, soil type and depth, and local socio-economic factors (Rao and Raju, 2010).

Rao and Raju (2010) had listed a set of parameters which need to be analyzed to fix appropriate locations for the water harvesting structures. These are:

- Rainfall.
- Land use or vegetation cover.
- Topography and terrain profile.
- Soil type & soil depth
- Hydrology and water resources.
- Socio-economic and infrastructure conditions.
- Environmental and ecological impacts.

Remote sensing techniques had been identified as potential tools to generate the basic information required for arriving at the most appropriate methods for each area.

In remote sensing aided analysis, various data layers were prepared and brought into a common GIS framework. Further, multi-criteria evaluation algorithms were used to aggregate the information from the basic data layers. Various decision rules were evaluated to arrive at the most appropriate solution as shown in the figure.

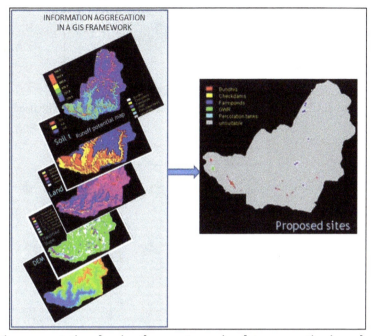

Schematic representation showing the remote sensing data aggregation in evaluating the suitability of various water harvesting techniques.

The capability to provide large areal coverage at a fine spatial resolution makes remote sensing techniques highly advantageous over the conventional field-based surveys.

Runoff Model

A runoff model is a mathematical model describing the rainfall–runoff relations of a rainfall *catchment area*, drainage basin or *watershed*. More precisely, it produces a surface runoff hydrograph in response to a rainfall event, represented by and input as a hyetograph. In other words, the model calculates the conversion of rainfall into runoff. A well known runoff model is the *linear reservoir*, but in practice it has limited applicability. The runoff model with a *non-linear reservoir* is more universally applicable, but still it holds only for catchments whose surface area is limited by the condition that the rainfall can be considered more or less uniformly distributed over the area. The maximum size of the watershed then depends on the rainfall characteristics of the region. When the study area is too large, it can be divided into sub-catchments and the various runoff hydrographs may be combined using flood routing techniques.

Rainfall-runoff models need to be calibrated before they can be used.

Linear Reservoir

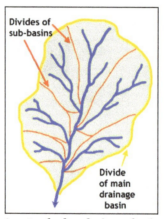

A watershed or drainage basin.

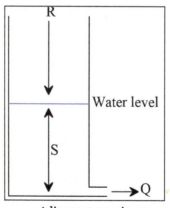

A linear reservoir.

The hydrology of a linear reservoir is governed by two equations.

1. flow equation: $Q = A.S$, with units [L/T], where L is length (e.g. mm) and T is time (e.g. h, day).

2. continuity or water balance equation: $R = Q + dS/dT$, with units [L/T].

where:

Q is the *runoff* or *discharge*.

R is the *effective rainfall* or *rainfall excess* or *recharge*.

A is the constant *reaction factor* or *response factor* with unit [1/T].

S is the water storage with unit [L].

dS is a differential or small increment of S.

dT is a differential or small increment of T.

Runoff Equation

A combination of the two previous equations results in a differential equation, whose solution is:

- $Q_2 = Q_1 \exp\{-A(T_2 - T_1)\} + R[1 - \exp\{-A(T_2 - T_1)\}]$

This is the *runoff equation* or *discharge equation*, where Q_1 and Q_2 are the values of Q at time T_1 and T_2 respectively while $T_2 - T_1$ is a small time step during which the recharge can be assumed constant.

Computing the Total Hydrograph

Provided the value of A is known, the *total hydrograph* can be obtained using a successive number of time steps and computing, with the *runoff equation*, the runoff at the end of each time step from the runoff at the end of the previous time step.

Unit Hydrograph

The discharge may also be expressed as: $Q = -dS/dT$. Substituting herein the expression of Q in equation (1) gives the differential equation $dS/dT = A.S$, of which the solution is: $S = \exp(-A.t)$. Replacing herein S by Q/A according to equation (1), it is obtained that: $Q = A \exp(-A.t)$. This is called the instantaneous unit hydrograph (IUH) because the Q herein equals Q_2 of the foregoing runoff equation using $R = 0$, and taking S as *unity* which makes Q_1 equal to A according to equation (1).

The availability of the foregoing *runoff equation* eliminates the necessity of calculating the *total hydrograph* by the summation of partial hydrographs using the *IUH* as is done with the more complicated convolution method.

Determining the Response Factor A

When the *response factor* A can be determined from the characteristics of the watershed (catchment area), the reservoir can be used as a *deterministic model* or *analytical model*, see hydrological modelling. Otherwise, the factor A can be determined from a data record of rainfall and runoff using the method explained below under *non-linear reservoir*. With this method the reservoir can be used as a black box model.

Conversions

1 mm/day corresponds to 10 m³/day per ha of the watershed.

1 l/s per ha corresponds to 8.64 mm/day or 86.4 m³/day per ha.

Non-linear Reservoir

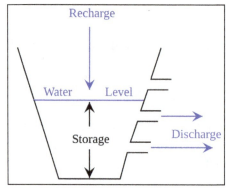
A non-linear reservoir.

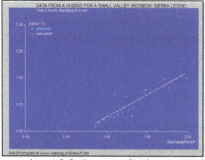
FThe reaction factor (Aq, Alpha) versus discharge (Q) for a small valley (Rogbom) in Sierra Leone.

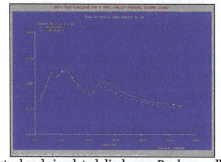
Actual and simulated discharge, Rogbom valley.

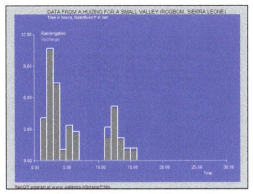
Rainfall and recharge, Rogbom valley.

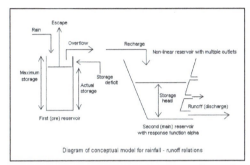
Non-linear reservoir with pre-reservoir for recharge.

Contrary to the linear reservoir, the non linear reservoir has a reaction factor A that is not a constant, but it is a function of S or Q.

Normally A increases with Q and S because the higher the water level is the higher the discharge capacity becomes. The factor is therefore called Aq instead of A. The non-linear reservoir has *no* usable unit hydrograph.

During periods without rainfall or recharge, i.e. when $R = 0$, the runoff equation reduces to:

- $Q_2 = Q_1 \exp\{-A_q(T_2 - T_1)\}$, or:

or, using a *unit time step* ($T_2 - T_1 = 1$) and solving for Aq:

- $A_q = -\ln(Q_2/Q_1)$

Hence, the reaction or response factor Aq can be determined from runoff or discharge measurements using *unit time steps* during dry spells, employing a numerical method.

The figure shows the relation between Aq (Alpha) and Q for a small valley (Rogbom) in Sierra Leone.

The figure shows observed and *simulated* or *reconstructed* discharge hydrograph of the watercourse at the downstream end of the same valley.

Recharge

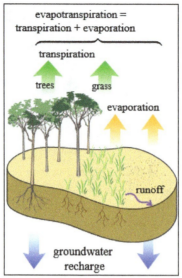

Runoff from the water balance.

The recharge, also called *effective rainfall* or *rainfall excess*, can be modeled by a *pre-reservoir* giving the recharge as *overflow*. The pre-reservoir knows the following elements:

- a maximum storage (Sm) with unit length [L].
- an actual storage (Sa) with unit [L].
- a relative storage: Sr = Sa/Sm.
- a maximum escape rate (Em) with units length/time [L/T]. It corresponds to the maximum rate of evaporation plus *percolation* and groundwater recharge, which will not take part in the runoff process.
- an actual escape rate: Ea = Sr.Em.
- a storage deficiency: Sd = Sm + Ea − Sa.

The recharge during a unit time step (T2−T1=1) can be found from R = Rain − Sd The actual storage at the end of a *unit time step* is found as Sa2 = Sa1 + Rain − R − Ea, where Sa1 is the actual storage at the start of the time step.

The Curve Number method (CN method) gives another way to calculate the recharge. The *initial abstraction* herein compares with Sm − Si, where Si is the initial value of Sa.

Software

Figures were made with the RainOff program, designed to analyse rainfall and runoff using the non-linear reservoir model with a pre-reservoir. The program also contains an example of the hydrograph of an agricultural subsurface drainage system for which the value of A can be obtained from the system's characteristics.

The SMART hydrological model includes agricultural subsurface drainage flow, in addition to soil and groundwater reservoirs, to simulate the flow path contributions to streamflow.

V*flo* is another software program for modeling runoff. V*flo* uses radar rainfall and GIS data to generate physics-based, distributed runoff simulation.

The WEAP (Water Evaluation And Planning) software platform models runoff and percolation from climate and land use data, using a choice of linear and non-linear reservoir models.

The RS MINERVE software platform simulates the formation of free surface run-off flow and its propagation in rivers or channels. The software is based on object-oriented programming and allows hydrologic and hydraulic modeling according to a semi-distributed conceptual scheme with different rainfall-runoff model such as HBV, GR4J, SAC-SMA or SOCONT.

SWAT Model

SWAT (Soil & Water Assessment Tool) is a river basin scale model developed to quantify the impact of land management practices in large, complex watersheds. SWAT is a public domain software enabled model actively supported by the USDA Agricultural Research Service at the Blackland Research & Extension Center in Temple, Texas, USA. It is a hydrology model with the following components: weather, surface runoff, return flow, percolation, evapotranspiration, transmission losses, pond and reservoir storage, crop growth and irrigation, groundwater flow, reach routing, nutrient and pesticide loading, and water transfer. SWAT can be considered a watershed hydrological transport model. This model is used worldwide and is continuously under development. As of July 2012, more than 1000 peer-reviewed articles have been published that document its various applications.

Model Operation

SWAT is a continuous time model that operates on a daily time step at basin scale. The objective of such a model is to predict the long-term impacts in large basins of management and also timing of agricultural practices within a year (i.e., crop rotations, planting and harvest dates, irrigation, fertilizer, and pesticide application rates and timing). It can be used to simulate at the basin scale water and nutrients cycle in landscapes whose dominant land use is agriculture. It can also help in assessing the environmental efficiency of best management practices and alternative management policies. SWAT uses a two-level dissagregation scheme; a preliminary subbasin identification is carried out based on topographic criteria, followed by further discretization using land use and soil type considerations. Areas with the same soil type and land use form a Hydrologic Response Unit (HRU), a basic computational unit assumed to be homogeneous in hydrologic response to land cover change.

Irrigation Management

Remote Sensing for Irrigation Management

In irrigation management, remote sensing is used as a tool to collect spatial and temporal variations in the hydro-meteorological parameters, crop characteristics and soil characteristics. Some of the important applications of remote sensing in irrigation management are listed below.

- Assessment of water availability in reservoirs for optimal management of water to meet the irrigation demand.
- Identifying, inventorying and assessment of irrigated crops.
- Determination of irrigation water demand over space and time.
- Distinguishing land irrigated by surface water bodies and by ground water withdrawals.
- Estimation of crop yield.
- Study on water logging and salinity problems in irrigated lands.
- Irrigation scheduling based on water availability and water demand.
- Evapotranspiration studies.
- Irrigation system performance evaluation.

Crop Classification and Identification of the Irrigated Areas

Crop classification using the satellite remote sensing images is one of the most common applications of remote sensing in agriculture and irrigation management. Multiple images corresponding to various cropping periods are generally used for this purpose. The spectral reflectance values observed in various bands of the images are related to specific crops with the help of ground truth data. Also, satellite images of frequent time intervals are used to capture the temporal variations in the spectral signature, using which the crop stages are identified. The table gives a sample list of spectral signatures, observed in the standard FCC, for different crops during different growth stages.

Table: Spectral signatures of different crops in different growth stages

Crop Type	Growth Stage at the time of Satellite Data Acquisition	Possible signature on a Standard FCC
Paddy	2 to 3 weeks after transplantation	Greenish black to Reddish black
Paddy	Peak vegetative phase	Dark Red
Groundnut	Peak vegetative phase	Shades of bright red
Sugarcane	Peak vegetative phase	Light Pink to Pink
Cotton	Peak vegetative phase	Pink to Red

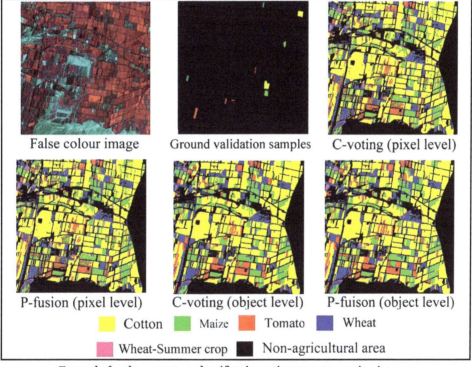

Example for the crop type classification using remote sensing images

Identification of the irrigated area from the satellite images is based on the assessment of the crop health (using vegetation indices such as Normalized Differential Vegetation Index, NDVI) and the soil moisture condition. Irrigation water demand of the crops is defined by the actual evapotranspiration (ET) and the soil moisture availability. Remote sensing application in the estimation of irrigation water demand employs the estimation of ET by using the plant bio-physical parameters and the atmospheric parameters, and the soil moisture condition. The table gives a list of crop bio-physical parameters and their application in irrigation management. The following table lists the capability of the remote sensing techniques in estimating these parameters.

Table: List of crop bio-physical parameters and their application in irrigation management.

Crop Parameter	Process	Purpose
Vegetation cover	Chorophyll development, soil and canopy fluxes	Irrigation area
Leaf area index	Biomass, minimum canopy resistance, heat fluxes	Yield, water use, water needs
Photosynthetically active radiation	Photosynthesis	Yield
Surface roughness	Aerodynamic resistance	Water use water needs
Surface albedo	Net radiation	Water use water needs
Thermal infrared surface emissivity	Net radiation	Water use water needs
Surface temperature	Net radiation, surface resistance	Water use
Surface resistance	Soil moisture and salinity	Water use
Crop coefficients	Grass evapotranspiration	Water needs
Transpiration coefficients	Potential soil and crop evaporation	Water use water needs
Crop yield	Accumulated biomass	Production

Table: Capability of the remote sensing techniques in estimating crop bio-physical parameters.

Parameter	Accuracy	Need for field Data
Vegetation cover	High	None
Leaf area index	Good	None
Photosynthetically active radiation	Good	None
Surface roughness momentum	High	None
Surface roughness heat	Low	High
Surface albedo	Good	Low
Thermal infrared surface emissivity	Good	None
Surface temperature	Good	Low
Surface resistance	Good	None
Crop coefficients: tabulated	Moderate	None
Crop coefficients: analytical	Moderate	High
Transpiration coefficients	Good	None

Performance of the irrigation system is generally evaluated using indices such as relative water supply and relative irrigation supply. Bastiaanssen et al. (1998) has listed a set of irrigation performance indices derived with the help of the remote sensing data. Soil-adjusted vegetation index (SAVI), NDVI, transformed vegetation index (TVI), normalized difference wetness index (NDWI), green vegetation index (GVI) are a few of them.

Flood Mapping

Role of Remote Sensing Data in Flood Analyses

Remote sensing facilitates the flood surveys by providing the much needed information for flood studies. Satellite images acquired in different spectral bands during, before and after a flood event can provide valuable information about flood occurrence, intensity and progress of flood inundation, river course and its spill channels, spurs and embankments affected/ threatened etc. so that appropriate flood relief and mitigation measures can be planned and executed in time.

Poor weather condition generally associated with the floods, and poor accessibility due to the flooded water makes the ground and aerial assessment of flood inundated areas a difficult task. Application of satellite remote sensing helps to overcome these limitations. Through the selection of appropriate sensors and platforms, remote sensing can provide accurate and timely estimation of flood inundation, flood damage and flood-prone areas.

A list of sensors used for flood analyses are given in the table below.

Table: List of satellite sensors with their use for flood monitoring (Bhanumurthy et al., 2010).

Sl No:	Satellite	Sensor/ Mode	Spatial Resolution(m)	Spectral Resolution (μm)	Swath (km)	Used For
1.	IRS-P6	AWiFS	56	B2 : 0.52-0.59 B3 : 0.62-0.68 B4 : 0.77-0.86 B5 : 1.55-1.70	740	Regional level flood mapping
2.	IRS-P6	LISS-III	23.5	B2 : 0.52-0.59 B3 : 0.62-0.68 B4 : 0.77-0.86 B5 : 1.55-1.70	141	District-level flood mapping
3.	IRS-P6	LISS-IV	5.8 at nadir	B2 : 0.52-0.59 B3 : 0.62-0.68 B4:0.77-0.86	23.9	Detailed Mapping
4.	IRS-1D	WiFS	188	B3: 0.62-0.68 B4 : 0.77-0.86	810	Regional level flood mapping
5.	IRS-1D	LISS-III	23.5	B2: 0.52-0.59 B3 : 0.62-0.68 B4 : 0.77-0.86 B5:1.55-1.70	141	Detailed Mapping
6.	Aqua / Terra	MODIS	250	36 in visible NIR & thermal	2330	Regional level Mapping

7.	IRS-P4	OCM	360	Eight narrow bands in visible & NIR	1420	Regional level Mapping
8.	Cartosat-1	PAN	2.5	0.5-0.85	30	Detailed Mapping
9.	Cartosat-2	PAN	1	0.45-0.85	9.6	Detailed Mapping
10.	Radarsat-1	SAR/ ScanSAR Wide	100	C-band (5.3 cm; HH Polarization)	500	Regional level mapping
11.	Radarsat-1	SAR /ScanSAR Narrow	50	C-band (5.3 cm)	300	District-level mapping
12	Radarsat-1	Standard	25	C-band	100	District-level mapping
13	Radarsat-1	Fine beam	8	C-band (5.3 cm)	50	Detailed mapping
14	Radarsat-2	SAR	3m ultra-find mode and 10m multi-llik fine mode	C –band	20 in ultra fine mode	Detailed mapping
14	ERS	SAR	25	C-band ; VV Polarization	100	District-level mapping

Remote Sensing Applications in Flood Analysis

The following 5 areas of remote sensing data application in flood analysis are identified:

- Flood mapping.
- Near real-time monitoring of floods.
- Flood damage assessment.
- Flood hazard mapping.
- River studies: mapping of river bank erosion and river course change.

Environmental Monitoring

Remote Sensing in Water Quality Monitoring

The term water quality indicates the physical, chemical and biological characteristics of water. Temperature, chlorophyll content, turbidity, clarity, total suspended solids (TSS), nutrients, colored dissolved organic matter (CDOM), tripton, dissolved oxygen, pH, biological oxygen demand (BOD), chemical oxygen demand (COD), total organic carbon, and bacteria content are some of the commonly used water quality parameters.

In remote sensing, water quality parameters are estimated by measuring changes in the optical properties of water caused by the presence of the contaminants. Therefore, optical remote sensing has been commonly used for estimating the water quality parameters.

The following figure shows the Landsat TM image of the Fitzroy Estuary and Keppal Bay in Australia. The image taken on May 2003 shows the color difference of the water near the estuary mouth, which is due to the presence of suspended sediments.

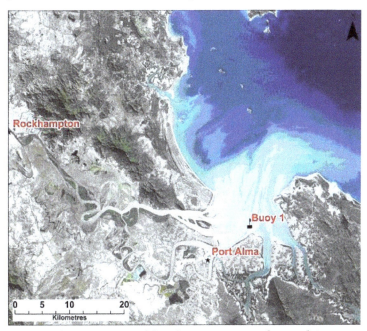

Landsat TM image of the Fitzroy Estuary and Keppal Bay in Australia in May 2003.

Water quality parameters that have been successfully extracted using remote sensing techniques include chlorophyll content, turbidity, secchi depth, total suspended solids, colored dissolved organic matter and tripton. Thermal pollution in lakes and estuaries is monitored using thermal remote sensing techniques.

Algorithms for the Estimation of Water Quality Parameters from Remote Sensing Data

Estimation of water quality parameters using remote sensing data is based on the relationship between the concentration of the pollutant in the water and the consequent changes in the optical properties as observed in the satellite image.

Wavelengths or Bands used for Water Quality Monitoring

Optimum wavelength for monitoring water quality parameter through remote sensing depends on the substance that is measured.

Based on several *in-situ* analyses, the VIS and NIR portions of the EMR spectrum with wavelengths ranging from 0.7 to 0.8 µm have been considered to be the most useful bands for monitoring suspended sediments in water.

Optical remote sensing using the VIS and NIR bands has been preferred for measuring Chlorophyl content, turbidity, CDOM, Tripton etc.

Algorithms used for the Estimation of Water Quality Parameters

Algorithms or models used for the estimation of water quality parameters can be classified into Empirical relationships, Radiative transfer models or Physical models.

Empirical models use the relationship between the water quality parameter and the spectral records. General forms of such relationships are the following (Schmugge et al., 2002):

$$Y = A + BX \quad or \quad Y = AB^x \quad (1)$$

where Y is the measurement obtained using the remote sensors and X is the water quality parameter of interest, and A and B are the empirical factors.

For example, an empirical relationship for estimating Chlorophyl content in water was given as follows (Harding et al., 1995):

$$Log_{10}[Chlorophyll] = A + B(-\log_{10} G) \quad (2)$$

$$G = \frac{(R_2)^2}{R_1.R_3} \quad (3)$$

where A and B are empirical constants derived from *in situ* measurements, R_1, R_2 and R_3 are the radiances at 460 nm, 490 nm and 520 nm, respectively.

Similarly, Eq. 4 shows the empirical relationship for TSS. The algorithm is used to detect the TSS in water using the MODIS data. It is also known as TSM Clark algorithm.

$$TSM = 10^{\sum_{i=0}^{5} a_i x^i}$$

Where $x = \log(nLw_1 + nLw_2 / nLw_4)$

and a_i = {0.490330, -2.712882, 3.412666, -8.336478, 12.111023, -5.961926}. (4)

where nLw1 and nLw2 and nLw4 are the normalized water-leaving radiances on the dark blue band, second blue band and green band, respectively. These are related to the subsurface irradiance reflectance R (For more details refer Brando and Decker, 2003). It is to be noted that TSS stands for Total Suspended Solids and TSM is an acronym for Total Suspended Matter. The equation (4) represents one of the TSM models which are empirical in nature. Relations such as equation (4) can also be developed for TSS using bands of MODIS imageries.

Such relationships, based on field observations of the water quality parameters and the corresponding measurements obtained using the sensor, are controlled by the properties of water such as density, temperature etc. Therefore, the relationship derived for one field condition may not be valid for the other areas

Radiative transfer models use a more general approach. Simplified solutions of the radiative transfer equations (RTEs) are used to relate the water surface reflectance (Rrs) to the controlling physical factors.

A sample RTE to relate the reflectance measured using remote sensing techniques to the suspended particulate matter is given below (Volpe et al., 2011):

$$R_{rs} = \frac{0.5 r_{rs}}{1 - 1.5 r_{rs}} \quad (5)$$

$$r_{rs} = r_{rs}^{dp}[1 - e^{-(K_d + K_u^C)H}] + \frac{\rho_b}{\pi} e^{-(K_d + K_u^B)H} \quad (6)$$

where,

r_{rs} = subsurface remote sensing reflectance.

r_{rs}^{dp} = r_{rs} for optically deep waters = $(0.084 + 0.17 u)u$.

u = $b_b / (a + b_b)$, where b_b is the backscattering coefficient and a is the absorption coefficient.

K_d = Vertically averaged diffuse attenuation coefficient for downwelling irradiance = $Dd\, \alpha$.

D_d = $1/\cos(\theta w)$, where θw is the subsurface solar zenith angle.

Ku^C = Vertically averaged diffuse attenuation coefficient for upwelling radiance from water column scattering = $Du^C \alpha$.

Ku^B = Vertically averaged diffuse attenuation coefficient for upwelling radiance from bottom reflectance = $Du^B \alpha$.

$\alpha = a + b_b$.

$Du^C = 1.03(1+2.4u)^{0.5}$.

$Du^B = 1.03(1+5.4u)^{0.5}$.

ρb = Bottom albedo.

H = water depth.

The backscattering and the absorption coefficients were determined by calibration.

Satellites and Sensors used for Water Quality Monitoring

Remote sensing of the water quality parameter in the earlier days employed fine resolution optical images from the satellites e.g., Landsat TM. However, poor temporal coverage of the images (once in 16 days) was a major limitation in such studies. With the development of new satellites and sensors, the spatial, temporal and radiometric resolutions have improved significantly. Using sensors such as MODIS (with 36 spectral bands) and MERIS (with 15 spectral bands) better accuracy in the estimation of water quality parameters is now possible.

A recent development in the remote sensing application in water quality monitoring is the use of hyper-spectral images in monitoring the water quality parameters. The large number of narrow spectral bands used in the hyper-spectral sensors help in improved detection of the contaminants and the organic matters present in water. Use of hyper-spectral images to monitor the tropic status of lakes and estuaries, assessment of total suspended matter and chlorophyll content in the surface water and bathymetric surveys are a few examples.

For more details on the hyperspectral remote sensing data application in water quality monitoring, refer Koponen et al., 2002; Thiemann and Kaufmann, 2002; Hakvoort et al., 2002; Lesser and Mobley, 2007.

The following table gives a brief summary of some of the works wherein the remote sensing data have been used for estimating the water quality parameters.

The figures show the application of remote sensing data for monitoring various water quality parameters.

Table: Important water quality parameters estimated and the characteristics of the sensors used (Source: Nagesh Kumar and Reshmidevi, 2013).

Parameter	Sensor type	Sensor / data	Remote sensing data characteristics	Algorithm used	Reference
Chlorophyll	MSS	MERIS	15 spectral bands, 300 m spatial resolution, poor temporal coverage	Spectral curves were calibrated using field observations	Koponen et al., 2002
				ESA BEAM tool box	Giardino et al., 2010
		Landsat TM	7 spectral bands, 30 m spatial resolution, poor temporal coverage	Empirical relation	Brezonik et al., 2005
		SeaWiFS, MODIS	More number of spectral bands, 250- 1000 m spatial resolution, better temporal coverage,	Band ratio algorithm	Lesht et al., 2013
	Hyperspectral	Hyperion	Better spectral resolution, 30 m spatial resolution, poor temporal coverage	Analytical method, Numerical radiative transfer model	Brando et al., 2003
				Bio-optical model	Santini et al., 2010
CODM, Tripton	Hyperspectral	Hyperion	Better spectral resolution, 30 m spatial resolution, poor temporal coverage	Analytical method, Numerical radiative transfer model	Brando et al., 2003
				Bio-optical model	Santini et al., 2010
Secchi depth, Turbidity	MSS	MERIS	15 spectral bands, 300 m spatial resolution, poor temporal coverage	Spectral curves were calibrated using field observations	Schmugge et al., 1992
				ESA BASE toolbox	Koponen et al., 2002
		Landsat TM	7 spectral bands, 30 m spatial resolution, poor temporal coverage	Empirical relation	Brezonik et al., 2005
TSS	MSS	MERIS	15 spectral bands, 300 m spatial resolution, poor temporal coverage	ESA BASE tool box	Giardino et al., 2010
		Landsat TM	7 spectral bands, 30 m spatial resolution, poor temporal coverage	Empirical relation	Brezonik et al., 2005

Surface temperature	Thermal	MODIS–LST	Better temporal coverage, 250-1000 m spatial resolution	MODIS Level-2 temperature data	Alcântara et al., 2010; Giardino et al., 2010
		AVHRR	5 bands (3 thermal bands), good temporal coverage, 1000-2000 m spatial resolution	Multi-Channel SST estimation algorithm (MCSST)	Politi et al., 2012

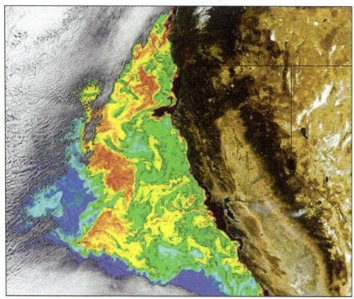

Chlorophyll concentration in the off-coast of California estimated using the SeaWiFS and MODIS sensors. Bright red indicates high concentration and blues indicate low concentrations.

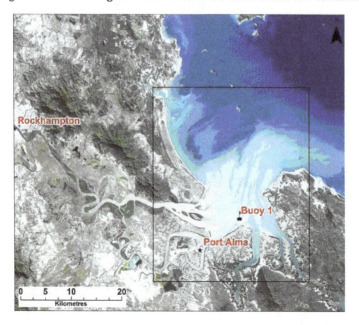

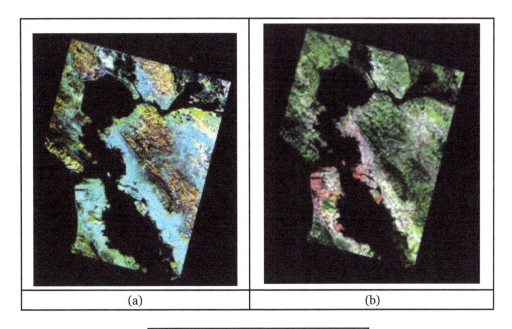
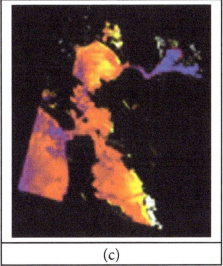

ASTER images of the San Francisco Bay area (a) From SWIR bands (b) A composite using thermal data and visible bands (c) Thermal data showing temperature variations only in water. Land areas are masked out.

In Fig.(c), colour varies from red for the warmest to blue for the coolest areas. The warmest temperatures are found in San Francisco and across the Bay in the Oakland group of cities, which may be mostly due to the thermal pollution from the large number of industries located in the area.

Remote Sensing Application in Monitoring Land Degradation

Land degradation is the deterioration of the land or soil properties that negatively affect the effective functioning of the land based ecosystems. From the agricultural perspective it may be defined as the reduction in soil capacity to produce crops. From the ecological perspective, land degradation causes damage to the healthy functioning of the land based ecosystems.

Land degradation may be either due to natural factors such as floods, drought, earthquake, or due to the human induced factors like over exploitation of land and water resources, or unscientific land use. The following are considered to be some of the major factors causing land degradation (Ravishankar and Sreenivas, 2010).

- Water erosion: Displacement of soil material by water.

- Wind erosion: Displacement of top soil by wind.

- Water logging: Extensive ponding for a long time affecting the productivity of the land.

- Salinization: Chemical imbalance in the soil causing desiccation of the plants or non- availability of essential nutrients to plants.

- Acidification: Increase in the hydrogen cations in the soil affecting the plant health.

- Anthropogenic: Mining, industries leading to decreased productivity of the land.

- Others: Barren areas, rocky waste areas, riverine sand areas, sea ingression areas etc.

Vast areas in the world are currently affected by land degradation. According to the Department of Land Resources, in 2005 around 55.27 million hectares of land in India is affected due to some sort of degradation . Scientific information about the degraded land, or rate of land degradation is necessary for land reclamation and management.

Permissions

All chapters in this book are published with permission under the Creative Commons Attribution Share Alike License or equivalent. Every chapter published in this book has been scrutinized by our experts. Their significance has been extensively debated. The topics covered herein carry significant information for a comprehensive understanding. They may even be implemented as practical applications or may be referred to as a beginning point for further studies.

We would like to thank the editorial team for lending their expertise to make the book truly unique. They have played a crucial role in the development of this book. Without their invaluable contributions this book wouldn't have been possible. They have made vital efforts to compile up to date information on the varied aspects of this subject to make this book a valuable addition to the collection of many professionals and students.

This book was conceptualized with the vision of imparting up-to-date and integrated information in this field. To ensure the same, a matchless editorial board was set up. Every individual on the board went through rigorous rounds of assessment to prove their worth. After which they invested a large part of their time researching and compiling the most relevant data for our readers.

The editorial board has been involved in producing this book since its inception. They have spent rigorous hours researching and exploring the diverse topics which have resulted in the successful publishing of this book. They have passed on their knowledge of decades through this book. To expedite this challenging task, the publisher supported the team at every step. A small team of assistant editors was also appointed to further simplify the editing procedure and attain best results for the readers.

Apart from the editorial board, the designing team has also invested a significant amount of their time in understanding the subject and creating the most relevant covers. They scrutinized every image to scout for the most suitable representation of the subject and create an appropriate cover for the book.

The publishing team has been an ardent support to the editorial, designing and production team. Their endless efforts to recruit the best for this project, has resulted in the accomplishment of this book. They are a veteran in the field of academics and their pool of knowledge is as vast as their experience in printing. Their expertise and guidance has proved useful at every step. Their uncompromising quality standards have made this book an exceptional effort. Their encouragement from time to time has been an inspiration for everyone.

The publisher and the editorial board hope that this book will prove to be a valuable piece of knowledge for students, practitioners and scholars across the globe.

Index

A
Accumulated Biomass, 224
Aerial Photography, 38-40, 42, 44, 60-61, 130, 143, 190
Aircraft Platform, 63
Amateur Radio, 175

G
Groundwater Recharge, 220

H
Helium, 23

I
Image Map, 145
Imaging System, 69-70
Impedance, 18
Information Systems, 139, 145, 156
Infrared Detector, 23
Integer, 78, 185-186
Intellectual Property, 172
Intellectual Property Laws, 172
Intermediate Frequency, 119-120, 124-125
Interoperability, 156
Interpolation, 79, 150-151
Intranet, 146
Inventory, 141
Irrigation, 26, 28, 207, 221-224
Irrigation Management, 207, 222-224
Irrigation Scheduling, 222

J
Javascript, 200
Jurisdiction, 154

L
Land Degradation, 233-234
Landscape Ecology, 133
Light Detection And Ranging, 67, 128-132, 135
Limiting Factor, 107
Linear Polynomial, 78

Logo, 159
Loops, 191
Low-pass Filter, 82

M
Mainframe, 141, 164, 170
Metadata, 156
Microprocessor, 103

N
National Security, 200
Natural Landscape, 61
Navigation Satellites, 160

O
Operating Systems, 142, 194
Organic Carbon, 227
Organic Matter, 227
Overlay Analysis, 75

P
Parameterization, 212
Percolation, 220-221
Photodiodes, 132
Photogrammetry, 1, 38-39, 145
Photon, 129
Polarization, 44
Policy Decisions, 156
Polynomial, 77-78
Polynomials, 78
Precipitation, 16, 210-211
Public Administration, 205
Pulse Duration, 13

Q
Quadratic Equation, 182
Quadratic Polynomial, 78
Quantitative Analysis, 143

R
Radio Waves, 102-105, 107-108, 110, 135, 145
Raster Data, 143-145, 149

Reaction Time, 126
Redundancy, 169
Remote Sensing, 1, 9, 11, 15, 19, 23, 27, 34, 38, 61, 65, 67, 69, 74, 80, 83, 97, 128, 135, 145, 207, 222
Remote Sensing Technology, 135
Resolution, 5, 11, 15-16, 23, 25, 34, 42, 71, 96, 110, 131, 135, 143, 147, 155, 177, 184, 190, 195, 199, 202, 204, 208-211, 215, 230
Reverse Geocoding, 151
Runoff, 148, 210, 215-221

S
Salinity, 15-16, 222, 224
Satellite Imagery, 8, 88, 92, 143, 190
Sea Level, 14, 48, 50, 55, 166
Semantic Web, 157
Semantics, 156-157
Semiconductor, 124
Signal Frequency, 118
Signal Processing, 102, 108
Social Sciences, 203
Soil Erosion, 207
Soil Moisture, 4, 15-16, 27-28, 70, 210, 212-214, 223
Spatial Analysis, 140-141, 146, 148
Spatial Data, 50, 139, 141-142, 145, 148, 150, 152-153

Spatial Information, 89, 139, 143, 155
Spatial Relationships, 145, 148
Spatial Scales, 14
Spectral Signature, 4, 6, 70, 222
Square Matrix, 17-18
Streamflow, 221
Surface Analysis, 156
Surface Runoff, 148, 215, 221
Suspended Solids, 227, 229
Syntax, 156

T
Tensile Strength, 1, 38, 102
Thematic Mapper, 23, 26, 68, 71, 152
Thematic Maps, 6-8, 143
Thermal Infrared, 8, 19, 23-27, 29, 35, 69-70
Time Scale, 166
Topography, 14, 27, 57, 78, 83, 98, 133, 138, 140, 190, 195, 209, 214
Topology, 87, 141

W
Web Browser, 154
Web Mapping, 155
World Wide Web, 142